Alicyclic Chemistry

Cambridge Chemistry Texts

GENERAL EDITORS

E. A. V. Ebsworth, Ph.D.
Professor of Inorganic Chemistry,
University of Edinburgh

D. T. Elmore, Ph.D.
Reader in Biochemistry,
The Queen's University of Belfast

P. J. Padley, Ph.D.
Lecturer in Physical Chemistry,
University College of Swansea

K. Schofield, D.Sc.
Reader in Organic Chemistry,
University of Exeter

Alicyclic Chemistry

F. J. McQUILLIN

Reader in Organic Chemistry,
University of Newcastle upon Tyne

CAMBRIDGE
at the University Press 1972

547.5
M 173

Published by the Syndics of the Cambridge University Press
Bentley House, 200 Euston Road, London NW1 2DB
American Branch: 32 East 57th Street, New York, N.Y.10022

© Cambridge University Press 1972

Library of Congress Catalogue Card Number: 75-176253

ISBN: 0 521 08216 1   Clothbound
      0 521 09659 6   Paperback

Printed in Great Britain by
William Clowes & Sons Limited, London, Beccles and Colchester

Contents

Preface

The study of alicyclic substances has proved a rich source of experimental information from which chemical principles of wide general significance have been derived. From early studies of the synthesis and chemical behaviour of stereoisomeric decalins and similar structures and the subsequent unfolding of sterol chemistry grew conformational analysis, a concept recognised by the award of a Nobel Prize to D. H. R. Barton and O. Hassell in 1969. The elaboration of rings of various sizes and of bridged and highly strained ring systems has not only stimulated the development of ingenious synthetic procedures, but has provided the basis for a further understanding of chemical bonding. To Baeyer's original concept of ring strain we can now give a more quantitative meaning in terms of bond deformations and bond–bond repulsion. The application of nuclear magnetic resonance spectroscopy has brought a new depth to our knowledge of chemistry, and in alicyclic chemistry n.m.r. study of conformational equilibria, of isomerisation, and of fluxional molecules and homoaromatic systems has brought notable advances. To alicyclic chemistry as a whole, however, the recent discussion of orbital symmetry correlations by R. B. Woodward and R. Hoffmann has added a new dimension of the greatest significance.

In this text I have attempted to illustrate these developments by means of simple examples but with sufficient references to the literature to enable a student to extend his reading. I have assumed that the reader will obtain from other sources an acquaintance with the principles of stereochemistry, spectroscopy and the general methods of chemical synthesis and these subjects are discussed only so far as is essential.

I am grateful to my colleague Dr R. J. Stoodley and to Dr K. Schofield for helpful criticism and suggestions. I am grateful also to Dr M. S. Baird who very kindly read the proofs.

F. J. McQuillin

1 Cycloalkanes, cycloalkenes and cycloalkynes

1.1. Introduction. Alicyclic chemistry is concerned with the chemical and physical properties of substances the molecular skeleton of which contains one or more groups of carbon atoms in a ring or cyclic structure, and which may contain alkane, alkene or alkyne units. Inclusion of such units in a ring necessarily limits the degree of rotational freedom about carbon–carbon bonds in comparison with the rotational freedom in analogous acyclic structures, and since the establishment of molecular conformations which minimise steric repulsions between proximate groups depends on rotational freedom, steric interactions are a particular feature of alicyclic chemistry. Steric interactions not only influence the energy of formation of cyclic structures but also determine the environment of functional groups and hence their reactivity. Ring formation may also, in particular cases, result either in compression or widening of the bond angles characteristic of a similar acyclic molecule which again modifies the reactivity of substituents as well as the thermodynamic stability of the structure. Alicyclic chemistry is therefore particularly concerned with correlations between structure, stability, and reactivity, and the study of reactivity has necessarily focused attention not only on conformational factors, but also on reaction mechanism and the nature of reaction transition states. In this context cyclic structures of known molecular geometry have provided a valuable means of examining the importance of the geometrical relation between reacting groups in determining the reaction pathway. Most recently the interpretation of a range of alicyclic cyclo-addition, electrocyclic, and sigmatropic reactions has led to the recognition of the wide significance also of molecular orbital symmetry correlations in determining structural stability and the course of many organic chemical reactions.

1.2. Nomenclature. In systematic nomenclature alicyclic compounds are distinguished by the prefix cyclo-, or where the structure comprises more than one closed ring of atoms by the prefix bicyclo-, tricyclo-, etc. The size of ring is indicated by use of the standard terminology for n-alkane, alkene, or alkyne chains of differing length. The nomenclature

1

for a series of cycloalkanes and cycloalkenes is illustrated in the following examples.

CH₃CH₂CH₃	CH₃(CH₂)₂CH₃	CH₃(CH₂)₃CH₃	CH₃(CH₂)₄CH₃
Propane	n-Butane	n-Pentane	n-Hexane

Cyclopropane	Cyclobutane	Cyclopentane	Cyclohexane

Cyclopropene	Cyclobutene	Cyclopentene	Cyclohexene

In the case of cycloalkenes containing several olefinic bonds the location of these groups may sometimes be obvious as in cyclopentadiene, cycloheptatriene or cyclo-octatetraene for which only one structure is possible. In other instances the location of the olefinic bonds is indicated by numbering the structure as for example: cyclo-octa-1,3-diene, or cyclohexa-1,4-diene:

Cyclopentadiene	Cycloheptatriene	Cyclo-octatetraene

Cyclo-octa-1,3-diene	Cyclohexa-1,4-diene

Other functional groupings, e.g. hydroxyl, carbonyl, carboxylic acid are indicated as in corresponding acyclic compounds, e.g.:

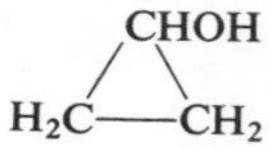

Cyclopropanol 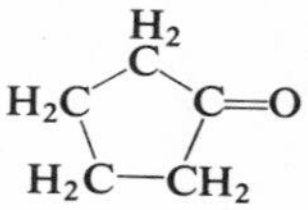   Cyclopentanone 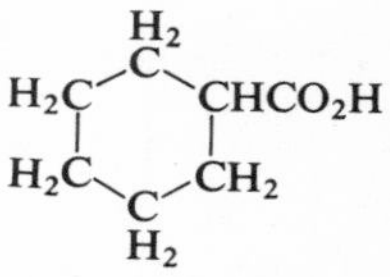   Cyclohexane-
                                                   carboxylic acid

With more than one functional group numbering is necessary to indicate their relationship, e.g.:

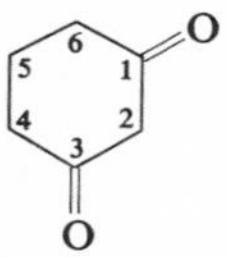

Cyclohexane-1,3-dione 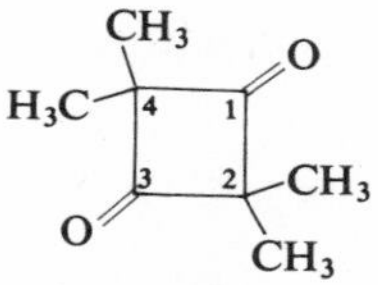   2,2,4,4-Tetramethyl
                            cyclobutane-1,3-dione

Where the functional groups occupy tetrahedrally substituted carbon atoms their steric relationship must also be specified.   In simple cases it is possible to use the prefix *cis-* or *trans-* as in the following examples.

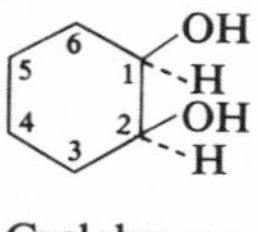

Cyclohexane   Cyclohexane
cis-1,2-diol   *trans*-1,2-diol

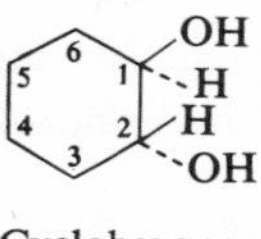

These structural formulae also indicate the use of a full line —, or a broken line -----, to designate bonds projecting up or down respectively, relative to the general plane of the molecule. With this convention the *cis* or *trans* relation of groups is made clear.

An alternative nomenclature for indicating the steric relation of groups has been applied especially within certain groups of natural products based on alicyclic structures, notably the sterols and terpenes. This makes use of the prefix β- or α- to indicate a bond projecting up or down

relative to the molecular plane. The use of this system is illustrated in
§2.12 and §9.3 and in the following example.

The system used in naming bridged and fused ring structures specifies,
in descending order, the number of carbon atoms which intervene between
the points of junction of the rings. The following illustrate the nomenclature applied to some fused ring systems:

Bicyclo[4,4,0]decane   Bicyclo[5,3,0]decane

In these instances the atoms common to the two rings are joined by
two chains of 4 and 4 or 5 and 3 atoms as indicated. The atoms (*a*) and
(*b*) being also joined directly the number of atoms in the third chain is
zero and this is specified. In the following examples of bridged rings
the third chain of atoms joining (*a*) and (*b*) is of one or more atoms and
this is indicated in the nomenclature.

Bicyclo[2,2,2]octane

Bicyclo[2,2,1]heptane

This system of nomenclature which is illustrated in a series of examples
in §8.1 may be extended to structures containing a number of rings.
Many alicyclic substances, however, have more convenient trivial
names. Thus bicyclo[2,2,1]heptane which represents the ring structure
of the camphor/borneol group of terpenes, without the terpenoid methyl

substituents, is known as norbornane. Bicyclo[4,4,0]decane is chemically

Bicyclo[2,2,1]heptane
= norbornane

Borneol

equivalent to decahydronaphthalene and is therefore generally described as decalin.

Naphthalene $\xrightarrow{10H}$ Decalin

Numbering of bicyclic substances commences from a ring junction and proceeds first along the longest chain to the next ring junction, then along the next longest and finally along the shortest, e.g.:

Bicyclo[2,2,1]heptane   Bicyclo[3,2,1]octane

However, for individual groups of substances other systems of numbering may be used. Thus the numbering in decalins is based on that used for naphthalene, whilst steroids or terpenes containing the decalin ring structure have their own numbering system.

Naphthalene   Decalin   Steroids

1.3. Cycloalkanes. The structural representation of ethane (1) expresses two important chemical concepts, the idea of directed bonding between atoms, and of repulsion between bonds. The bonding force

directed along the carbon–carbon and carbon–hydrogen internuclear axes establishes the four co-ordinate, near tetrahedral, steric arrangement around each carbon atom. Repulsion between adjacent bonds and between non-bonded atoms determines what is energetically the optimum conformation of the molecule. For ethane, bond repulsion is minimised in the staggered arrangement of the carbon–hydrogen bonds indicated in (1), or, as viewed along the carbon–carbon bond, in the Newman projection. By rotation about the carbon–carbon bond through 60°, this staggered arrangement (1) is converted into (2), in which the carbon–hydrogen bonds are mutually eclipsed as is indicated in the Newman projection. The bond–bond repulsion introduced by eclipsing presents a barrier of 11.8 kJ/mole to free rotation in ethane, that is bond–bond repulsion raises the energy of the eclipsed conformation (2) some 11.8 kJ/mole above that of the staggered arrangement (1).

For propane the eclipsed conformation (4) represents an energy level 14 kJ/mole greater than that of the staggered arrangement (3).

For butane there are two staggered conformations, the *anti*-arrangement (5), in which the two CH_3 groups are at maximal separation, and the gauche conformation (6), where the two $C-CH_3$ bonds subtend an angle of 60°. The gauche form (6) represents an energy of 3.3 kJ/mole above that of (5).

There are also two eclipsed conformations for butane, viz. (7) and (8) which correspond, respectively, with an energy level of ~14 and ~24 kJ/mole above the energy minimum of the *anti*-conformation (5).

(7) (8)

(9) (10) (11) (12)

When a CH_2—CH_2 unit is incorporated into a carbocyclic ring as in (9) the balance of forces which establishes the ethane conformation (1) will be modified, and in a manner depending to a large extent on the size of the ring.

1.4. Ring strain. The smallest rings as in cyclopropane (10) or cyclo-butane (11) are associated with the classical concept of angle strain as a consequence of compression of the normal alkane carbon–carbon bond angle. A much more general consequence of cyclisation, however, is the restriction of internal rotational freedom so that a cyclic molecule may be impeded in establishing the minimal energy state of a staggered arrange-ment of bonds as in ethane. The consequent unbalanced bond–bond repulsion forces introduce into the molecule torsional strain leading to bond deformation by bending, i.e. to altering bond angles inside or out-side the ring in comparison with corresponding bond angles in an acyclic alkane chain.

The characteristic tetrahedral arrangement of bonds about an alkane carbon atom is formally derived from the hybridisation of the directed 2p-orbitals with the spherically symmetrical 2s-orbital of the carbon atom. Equal mixing of the three 2p- and one 2s-orbital gives the (ideally) symmetrical tetrahedral sp^3-arrangement of bonds. Deformation of this symmetry may be expresssed as an alteration in the relative s/p-character of the hybrid orbital forming a bond. Therefore angle deformation in a cyclic structure may be expressed by describing the hybrid character of individual bonds as an $s + \lambda p$ mixture where λ is an appropriate mixing coefficient (Mislow, 1965). This convention does not alter the existence of ring-strain, but provides a convenient means of interpreting the relationship of ring size to bond properties and chemical reactions with which alicyclic chemistry is concerned.

1.5. The conformation of rings. In the smallest cycloalkanes, cyclopropane (10), cyclobutane (11), and cyclopentane (12) the carbon–carbon bond directions establish an essentially planar structure. Adjacent CH_2 units are, however, thereby brought into a eclipsed relation and the consequent torsion leads to some twisting about the plane of the ring.

In addition to C–H and C–C opposition forces there is, in the smaller rings, also what is described as angle strain due to compression of the carbon–carbon bond angle which falls: cyclopropane > cyclobutane > cyclopentane. The formal ring angles of 60° and 90°, however, do not represent the true carbon–carbon bond angles in cyclopropane or cyclobutane; this will be discussed below.

Cyclohexane forms a non-planar or puckered ring of tetrahedral

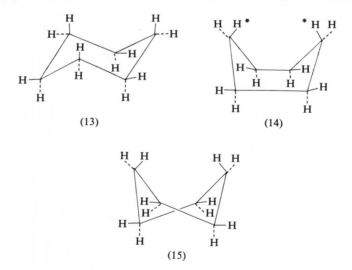

(13)   (14)

(15)

carbon units free of angle strain, which in the chair conformation (13) is composed of gauche butane units, with a minimum of bond opposition strain. For cyclohexane an alternative conformation, also free of angle strain, is the boat conformation (14). This boat conformation (14) contains eclipsed butane units and also a transannular interaction between the hydrogen atoms marked H* which approach to 180 pm, i.e. within the sum of their Van der Waals radii (240 pm; formerly 2.4 Å). However, the conformation (14) is flexible and these compressions may be minimised by flexing into the twisted boat conformation (15). The twisted boat conformation, however, still retains bond opposition forces amounting to *c*. 23 kJ/mole more than the chair conformation (13).

Infrared and Raman spectra, electron diffraction, and calculations of heat capacity indicate that cyclohexane and its derivatives adopt the chair conformation (13) wherever possible.

The energetically favourable arrangement represented by the chair conformation of cyclohexane is apparent in the heats of combustion per CH_2 group of the cycloalkanes (Table 1.1) which show a minimum at cyclohexane (cf. Eliel, 1962). The difference term $\Delta = H/n - 661.1$

TABLE 1.1   *Heats of combustion H* (kJ/mole) *per* CH_2 *group for cyclo-alkanes* $(CH_2)_n$ *as the vapour*

n	H/n	Δ	n	H/n	Δ
3	699.7	38.6	9	666.9	5.8
4	688.5	27.4	10	666.1	5.0
5	666.5	5.4	11	664.3	4.2
6	661.1	—	12	662.3	1.2
7	664.9	3.8	13	662.7	1.6
8	666.1	5.0	14	661.1	0

kJ/mole, comparing the enthalpy per CH_2 group of the cycloalkanes with the value for cyclohexane is very informative. For cyclopentane the total strain energy of $5 \times 5.4 = 27$ kJ/mole is appreciably smaller than would be expected having regard to the bond opposition torsional energy due to five eclipsed CH_2 groups of *c*. 40 kJ/mole. The difference (13 kJ/mole) is a measure of the relief in compression due to twisting of the cyclopentane ring which is therefore represented as having the envelope conformation (16) or the half-chair conformation (17). For cyclopropane

(16)   (17)

and cyclobutane the enthalpy difference indicates strain energies of 116 and 110 kJ/mole respectively, which are considerably larger than the strain due to bond opposition. The difference is ascribed to angle strain.

Cyclopropane, cyclobutane, cyclopentane, and the chair form of cyclohexane have very limited flexibility, i.e. they form essentially rigid structures. The larger cycloalkanes, however, have appreciable flexibility of the kind observed in the twisted boat form of cyclohexane. There is therefore in these cases uncertainty regarding the optimum conformation, or indeed whether a conformational equilibrium between different forms of closely similar energy may not offer the best account of the observed properties.

Cycloheptane and higher cycloalkanes may like cyclohexane be constructed from closely tetrahedral carbon units free from angle strain. The enthalpy difference term in Table 1.1 of 4–5 kJ/mole per CH_2 group over the range from cycloheptane to cycloundecane arises from bond opposition forces due to incomplete staggering of adjacent bonds, or to transannular interaction between substituents across the ring. The nature of these interactions may be considered in relation to cyclooctane. Here three main conformations have to be considered: the

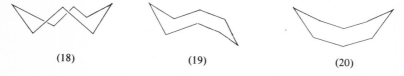

(18)   (19)   (20)

crown (18), the boat-chair (19), and the saddle-conformation (20). The most reliable evidence: n.m.r. data (Anet and Jacques, 1966) and X-ray analysis (on cyclo-octane-*trans*-1,2-dicarboxylic acid or *trans*-1,4-dichloro-cyclo-octane) (c.f. Dunitz *et al.* 1961), point to the boat-chair (19) as the preferred form, but an equilibrium involving this and a stretched crown conformation derived from (18) may be involved. However, neither in the boat-chair nor the crown conformation is complete staggering of bonds possible. There is therefore some bond opposition or torsional strain. It is important also to note the proximity of the pair of hydrogen substituents on atoms 1 and 5, cf. (21) which are brought together inside the ring. This transannular interaction effect, which has already been noted, is a feature of the medium-sized rings: $(CH_2)_n$, $n = 8$ to 11 (cf. chap. 3). Transannular interaction between 'turned inside' substituents is seen also in cyclodecane for which the preferred conformation is based on a deformed type of crown structure (22).

Detailed X-ray analysis, on *trans*-1,6-diaminocyclodecane hydro-chloride by Dunitz and Mugnoli (1966) shows, however, that the bond interactions in conformation (22) are in fact reduced by expansion of the carbon–carbon bond angles to an average of some 117°, i.e. above the normal paraffinic angle of 112°, and some angles to as much as 120°. The important transannular interaction of two sets of three hydrogen atoms is indicated in (22). However, in cyclododecane the carbon–carbon bond angle has the normal alkane value of *c*. 112°, and the separation of the nearest transannular hydrogen atoms is now near the sum of the Van der Waals radii.

(21)

(22)

(23)

1.6. Bond bending and stretching. The energy E of deforming a carbon bond angle by θ radians is given by:

$$E = \tfrac{1}{2}k(\theta)^2,$$

where k is the force constant, or for small angles, approximately $0.74\ \theta^2$ kJ/mole deg^2, amounting to about 2 kJ/mole for a deformation of 5° (Eliel, 1962; Mislow, 1965).

Torsional energy E_T due to bond opposition forces decreases with angular displacement (ω) from an eclipsed conformation:

$$E_T = \tfrac{1}{2}V \quad (1 + \cos n\omega),$$

where V is the height of the rotational energy barrier, in ethane ~ 12 kJ/mole, $\omega =$ the angular displacement from the eclipsed situation, and n, which for an alkane $= 3$, is the frequency of repetition of the same con-formation in a complete revolution. Clearly, bond opposition strains may be appreciably reduced by relatively small expenditure of energy in opening suitable bond angles. In the result, the energy of widening the bond angles in cyclodecane, for example is compensated by the consequent fall in torsional strain energy. Bond stretching, on the other hand, requires considerable energy; Table 1.2 compares some bending and

TABLE 1.2   *Bond bending and stretching force constants*

Bond	k/2 stretching (kJ/mole m² × 10⁻²⁰)	Angle	k/2 bending (kJ/mole rad²)
C—H	1450	C—C—H	164
C—C	1360	C—C—C	242
		H—C—H	97

stretching force constants (Eliel, 1965). Deforming a single C–C or C–H bond by as little as 30 pm requires about 1.2 kJ/mole. Bond stretching does not therefore contribute significantly to minimising strain energy of cyclic systems. Conversely bond angle deformation is very important in reducing strain, and consequently in altering the hybrid character of bonds, also their length, strength, and reactivity.

1.7. Bond angle and hybridisation. Table 1.3 summarises the change in bond properties with the state of hybridisation of carbon in the series ethane, ethylene and acetylene. Increasing *s*-character in the series of hybrids sp³, sp², sp is reflected in shorter and stronger C–H and C–C

TABLE 1.3

	CH_3—CH_3	CH_2=CH_2	CH≡CH
Length C—C (pm)	154	134	120
Length C—H (pm)	110.3	108.6	105.7
CH dissociation energy (kJ/mole)	409	422	460
ν_{CH}(cm⁻¹)	2960, 2860	3010–3060	3300
Angle H–C–C	110°	120°	180°

bonds, and consequently a higher ν_{CH} infrared stretching frequency. The principle underlying these changes in bond properties with hybridisation is related to the behaviour of strained alicyclic rings. For a particular $-CH_2-$ element in a ring, cf. (23) alteration of the carbon–carbon bond angle β, and hence the hybrid character of the orbitals forming the C–C bonds, will cause a compensating change in the carbon–hydrogen bond angle α, and in consequence the hybridisation and strength of the C–H bonds. This effect is illustrated by the experimental data for cyclopropane summarised in (24). The H–C–H bond angle is wider than in an acyclic alkane or strainless cycloalkane, and the C–H bond

α	118°
C—C	154 pm
C—H	108 pm
ν_{CH}	3010 cm^{-1}

(24)

length and stretching frequency are more nearly characteristic of the values for an sp^2 hybrid bond than for an alkane (cf. Table 1.3). This influence of the ring environment on the strength of external bonds is seen in a more general manner in the effect of ring size on the carbonyl group stretching frequency shown in Table 1.4, and on the C–Br bond dipole moment in Table 1.5. These data may be considered in terms of changing electronegativity of the ring carbon atoms towards external substituents; the carbonyl stretching bands for cyclopropanone and cyclobutanone occur in the same spectral region as ν_{CO} for acid halides or anhydrides, i.e. as carbonyls bearing strongly electronegative substituents.

TABLE 1.4

ν_{CO} for cycloalkanones		$O{=}C\ (CH_2)_{n-1}$	
n	$\nu_{CO}(cm^{-1})$		
3	1815	8	1692
4	1780	9	1698
5	1746	10	1694
6	1715	11	1700
7	1705	Acetone	1720

TABLE 1.5   C–Br *bond dipole moment in* BrCH(CH$_2$)$_{n-1}$

n	μ, Debye	n	μ, Debye
3	1.69	5	2.16
4	2.09	6	2.31

isopropyl bromide, μ 2.05 D

The data of Tables 1.4 (Ferguson, 1963) and 1.5 (Roberts and Chambers, 1951) provide a general basis for relating decreasing ring angle with increasing strength of external bonds. The increased bond strength may be expressed as due to greater s-character in the atomic orbital engaged; it is well known that the carbon 2s orbital represents an energy level some 512 kJ/mole below the 2p orbitals.

The angle θ between similar bonds $C\diagup^X_{\diagdown X}$ is related to the bond character by the expression:

$$1 + \lambda^2 \cos \theta = 0,$$

where λ^2 is the index of p-character in the hybrid (Mislow, 1965). When θ = the tetrahedral angle, $\lambda^2 = 3$, and the bond hybrid is sp^3. When $\theta = 180°, \lambda^2 = 1$ corresponding to the *sp* hybrid of acetylene. It is, however, clear that although the orbitals bonding individual groups to a central carbon atom may differ in relative s/p character the total s orbital contribution over all the bonds remains unity. Thus in cyclopropane, for example, a relatively greater s-character in the C–H bonds is balanced by a greater p-component in the C–C bonds of the ring. There is therefore no shortening of the C–C bonds, which thermally (cf. Table 1.1), and in their ease of rupture in chemical reactions, are evidently rather weaker than the C–C bonds of an alkane or of a less-strained carbocyclic. The observed carbon–hydrogen bond angle in cyclopropane (118°) nevertheless represents a surprisingly small deformation from the tetra-hedral† if the effective carbon–carbon ring angle is regarded as the inter-nuclear angle of 60°. Bonding by orbital overlap is not, however, necessarily directed along the internuclear axis. In ethylene, whether it is represented with bent bonds (25) or with a two-component σ/π bond (26), there is some measure of lateral overlap of orbitals. Cyclopropane

† i.e. the near tetrahedral angle of an alkane. Only in fully symmetrical compounds of the type CX$_4$ can the angle be truly tetrahedral

may similarly be represented using bent bonds which overlap as in (27).‡

$$(25) \qquad\qquad (26)$$

$$(27)$$

Calculations based on a model with the C–C bonding orbitals directed as in (27) give very satisfactory agreement with the properties of cyclopropane if the interorbital angle θ_o is taken as between 104° and 109°. Similar calculations for cyclobutane indicate an interorbital angle of $\sim 111.5°$.

1.8. Hybridisation and ^{13}C–H coupling. The proton spin–spin coupling constant in n.m.r. spectra is well known to be a function of the local electronic environment. For coupling between hydrogen and ^{13}C-isotopic carbon the change in $J_{13_{CH}}$ with ring size noted in Table 1.6 is clearly related to changing ^{13}C–H bond strength (cf. Jackman and

TABLE 1.6

$J_{13_{CH}}$ values for:		$\begin{array}{c} H \\ \text{C}^{13} \\ H \end{array} (CH_2)_n$	
n	Hz	n	^{13}CH Hz
2	161	6	123
3	134	9	122
4	128	11	118
5	124		

Sternhell, 1969). The data parallel the correlation of ring size, carbon–carbon bond angle, and state of hybridisation noted above. J_{13CH} has

‡ or, more correctly, by use of the molecular orbitals (Walsh, 1949; Hoffmann, 1968)

been empirically correlated with ring geometry, using, e.g. for cyclopropane a model with a carbon–carbon interorbital angle of 105.5° corresponding with a C–C bond $sp^{3 \cdot 7}$ hybrid, and a C–H bond hybrid corresponding to $sp^{2 \cdot 5}$.

1.9. Carbon acids.   These deductions are relevant to the strength of cyclopropane as a carbon acid: $\geqslant$C—H + Base→$\geqslant$C$^-$ + BH$^+$. The data of Table 1.7 place cyclopropane between an olefin and an alkane in

TABLE 1.7

	pK_a		pK_a
Ethane	42	Cyclopropane	39
Ethylene	36.5	Cyclobutane	43
Acetylene	25	Cyclopentane	44
		Cyclohexane	45

relative ease of proton removal (Cram, 1965). This emphasises the relative stability of the cyclopropyl carbanion (28), i.e. the strength of binding of this electron pair to the carbon nucleus.

(28)

1.10. Ultraviolet absorption of cyclopropanes.   The relatively strained C–C bonds in cyclopropane, on the other hand, correspond with rather less effective binding of the electrons in these bonds. Substances containing the cyclopropane ring show ultraviolet absorption in the region 190–200 nm (cf. propane λ_{max} 135 nm, ethylene λ_{max} 171 nm), i.e. the cyclopropane ring can engage in conjugation. This conjugative effect (cf. Kosower and Ito, 1962) is seen in the spectra of (29) and (30) and of (32) in comparison with (31). In the series (33), (34) and (35) the conjugative effect is clearly evident in the values of λ_{max}. In (33) the spiran structure holds the C–C bonds of the cyclopropane residue in the plane of the *p*-π-orbitals of the enone system and hence maintains the cyclopropane in the relation necessary for conjugative overlap. In (34) and

(35) the λ_{max} values show that other cycloalkyl groupings do not enter into a conjugative interaction of this kind.

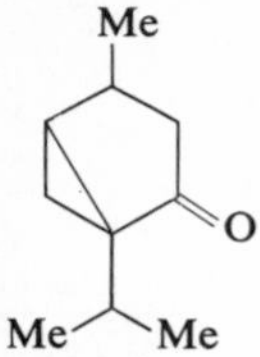

(29)

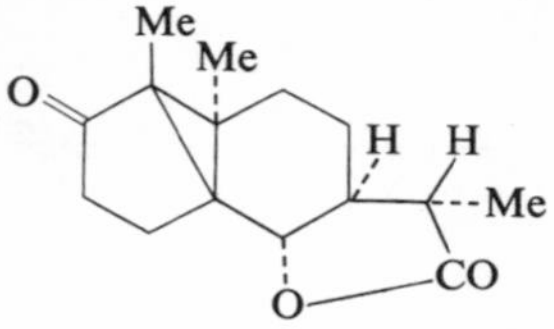

(30)

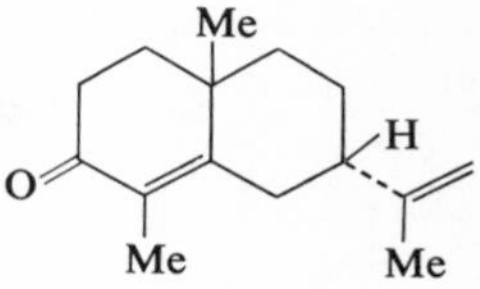

(31)

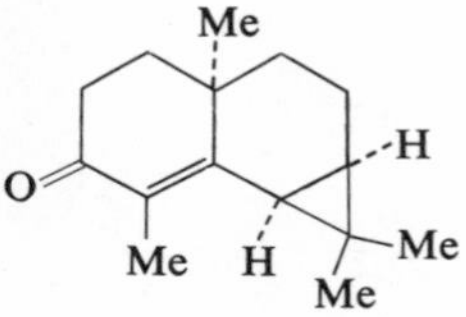

(32)

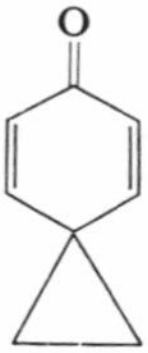

(33)

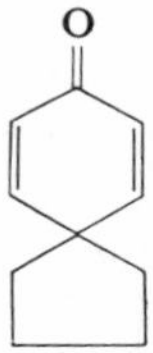

(34)

(35)

(29)
λ_{max} 210 nm, ϵ 2470
dihydroumbellulone

(30)
λ_{max} 214 nm, ϵ 4600
dihydrolumisantonin

(31)
λ_{max} 241 nm, ϵ 16 000

(32)
λ_{max} 266 nm, ϵ 13 700

(33)
λ_{max} 274
ϵ 21 900

(34)
242
15 900

(35)
242.5 nm
14 800

1.11. Proton chemical shifts in cycloalkanes. Chemical shift values (τ relative to TMS at 60 MHz) for the protons of cycloalkanes:

Cyclopropane 9.78, Cyclobutane 8.03,
Cyclopentane 8.49, Cyclohexane 8.57,
Cycloheptane 8.47, Cyclo-octane 8.47

indicate the marked proton shielding introduced by the electronic nature of the cyclopropane ring.

1.12. Cycloalkenes. Replacement of a –CH$_2$—CH$_2$– unit in a cyclo-alkane by an olefinic group –CH=CH– brings into the system a slightly shorter C–C distance, a larger carbon–carbon–carbon bond angle, and

TABLE 1.8   *Heats of hydrogenation* $-\Delta H$ (kJ/ mole) *for the cycloalkenes*

	(i) *cis*-cycloalkenes				
n	$-\Delta H$	Δ	n	$-\Delta H$	Δ
3	226.4	+108.4	8	96.6	−21.4
5	110.0	−8.0	9	99.1	−18.9
6	118.0	—	10	86.5	−31.5
7	111.3	−6.7			

Δ = value compared with cyclohexene

	(ii) *trans*-cycloalkenes		
n	8	9	10
$-\Delta H$	134.4	113.3	101.2

The olefinic CH stretching frequency of the following *cis*-cycloalkenes is also indicative of varying ring strain:

$\nu_{CH}(cm^{-1})$ for

n	cm^{-1}	n	cm^{-1}	n	cm^{-1}
3	3076	5	3061	7	3020
4	3048	6	3024	8	3016
		norbornene 3070 cm^{-1}			

the torsional rigidity of the C=C bond. There are, on the other hand, two fewer hydrogen or other substituent groups, i.e. fewer bond oppositions. In the smallest rings the wider bond angle at the trigonal carbon atom will increase ring strain, but in the medium-sized cycloalkenes this wider angle conforms to the requirements of a medium-sized ring.

The effect of introducing an olefinic group into a cycloalkane is reflected in the relative heats of hydrogenation, i.e. the enthalpy of the process:

The data in Table 1.8 (Eliel, 1962) indicate the anticipated angle strain in cyclopropene, but also illustrate the importance of bond opposition forces in the cycloalkane hydrogenation product from the five-, seven-, eight-, nine-, and ten-membered cycloalkenes, since for these cycloalkenes the heats of hydrogenation are low compared with that of cyclohexene. The relatively high ν_{CH} value for norbornene arises from the bending of the cyclopentene ring due to bridging, cf. (36). Similarly the heat of hydrogenation of norbornene, i.e. 139 kJ/mole is 29 kJ/mole greater than for cyclopentene.

(36)

In eight-membered and larger rings cycloalkenes are known in the *trans* as well as the *cis* form. In the eight- to ten-membered rings the *trans* olefinic bond, however, introduces greater angle strain, which is apparent from the relatively higher heat of hydrogenation (cf. Table 1.8). In eleven- and twelve-membered rings the *cis* and *trans* isomers are of similar energy and in larger rings the *trans* form becomes considerably the more stable, as in acyclic olefins. Acid catalysed equilibration:

which has been studied, gave the values of the equilibrium constants in Table 1.9 (Cope *et al.*, 1959).

TABLE 1.9 *Equilibration of cis- ⇌ trans-cyclo- alkenes using naphthalene-2-sulphonic acid at 100.4°C*

Ring size	9	10	11	12
$K_{cis/trans}$	232	12.2	0.406	0.517

1.13. Optically active *trans*-**cycloalkenes.** *trans*-Cyclo-octene is a dissymmetric molecule, cf. (37a) and (37b) and since the molecular packing prevents rotation of the ring of methylene groups around the olefinic bond, reasonably stable optically active forms are possible. Resolution of *trans*-cyclo-octene was achieved by Cope *et al.*, (1963) *via* the platinum chloride complex (38) carrying an optically active

α-phenylethylamine ligand. Crystallisation separated the complexes of (+)- and (−)-*trans*-cyclo-octene from which the (+)- and (−)-cyclo-alkenes were released by the action of potassium cyanide.

trans-Cyclononene has been resolved in the same way, but racemises more easily (Cope *et al.*, 1965). In *trans*-cyclodecene the olefinic group can rotate inside the loop of the methylene groups, and optical resolution has not been possible. The activation barrier for thermal racemisation has been found to be 153 and 84 kJ/mole for *trans*-cyclo-octene and *trans*-cyclononene respectively. The corresponding barrier in *trans*-cyclo-decene has been estimated to be some 42 kJ/mole from the temperature change in the n.m.r. signal of hexadeutero-*trans*-cyclodecene.

1.14. Relative reactivity of cycloalkenes. Some indication is available of the relative reactivities of the cycloalkenes (cf. Patai, 1964). Table 1.10 lists the rate constants for addition of di-isoamylborine. The sequence in Table 1.10 has an interesting correspondence with the equilibrium constants for complexing with silver ion in Table 1.11.

The ready complexing of more strained olefins with silver ion is to be noted (Table 1.11), cf. Mühs and Weiss (1962), and is of practical

TABLE 1.10 $k_2 \times 10^{-4}$ 1/mol s *for reaction of cycloalkenes with di-isoamylborine at* 0°C

| Cyclopentene | 14 | Cycloheptene | 266 |
| Cyclohexene | 0.13 | *cis*-Cyclo-octene | 580 |

value. *trans*-Cyclo-octene was purified *via* the silver nitrate adduct from which it was regenerated by the action of ammonia (Cope *et al.*, 1963).

TABLE 1.11   *Equilibrium constants for complexing with AgNO₃ in diethylene glycol at 37°C*

Cyclopentene	7.3	*cis*-Cyclo-octene	14.4
Cyclohexene	3.6	*trans*-Cyclo-octene	10^3
Cycloheptene	12.7	Norbornene	62

Very strained cyclo-olefins such as cyclopropene may polymerise, or as in the case of *cis, trans*-1,5-cyclo-octadiene may dimerise (39) → (40):

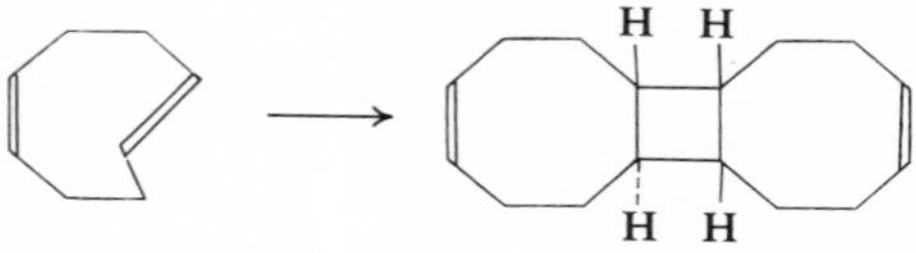

(39)   (40)

Strained cyclo-olefins are also characterised by relatively rapid reactions with a variety of dipolar addends, e.g. phenyl azide:

1.15. Cycloalkynes. Cycloalkynes with eight-membered and larger rings are known as reasonably stable substances, and in five-, six-, and seven-membered rings there is evidence for the existence of cyclo-alkynes as transient reaction intermediates (Blomquist *et al.*, 1952, 1953; Cope *et al.*, 1960; Roberts *et al.*, 1965; Wittig, 1960).

Curtius oxidation of the bishydrazone of a 1,2-dione:

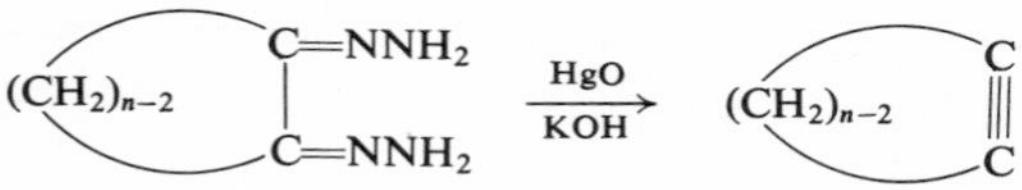

has been successfully applied to the cases $n = 5, 8, 9, 10$. The small ring cycloalkynes are trapped as diene adducts e.g.:

Formation of the cycloalkyne by dehydrohalogenation of a vinyl halide may be illustrated by the instance:

$$* = {}^{14}C$$

the ${}^{14}C$ labelling in the benzoic acid obtained by oxidation indicating the course of the reaction.

Catalytic hydrogenation of the cycloalkyne leads to the *cis*-cyclo-alkene. However, in eight-, nine-, and ten-membered cycloalkynes, Birch-reduction is anomalous in also giving the *cis*- rather than the *trans*-cycloalkene as the major product (Sicher *et al.*, 1964). An allene is generally thermally unstable relative to the corresponding in-chain acetylene. However, the enthalpy difference is not large, and in certain medium-sized rings the stabilities are reversed. Sodium amide, produced under the conditions of the Birch-reduction, provides a base for the interconversion:

With sodium amide in liquid ammonia as the base catalyst the following allene/acetylene ratios were found after the equilibration:

Ring size	9	10	11
Allene/Acetylene	20	1	0.1

Since the allene is reduced more rapidly, reaction takes the course:

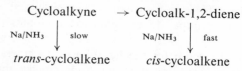

Cycloalkyne → Cycloalk-1,2-diene

Na/NH₃ | slow   Na/NH₃ | fast

trans-cycloalkene   *cis*-cycloalkene

leading to the *cis*-alkene as the main product.

References

Anet, F. A. L. and Jacques, M. St (1966). *J. Amer. Chem. Soc.*, **88**, 2585; 2586.
Blomquist, A. T., Liu, L. H. and Bohrer, J. C. (1952). *J. Amer. Chem. Soc.*, **74**, 3643, cf. Blomquist, A. T. and Liu, L. H. *J. Amer. Chem. Soc.*, 1953, **75**, 2153.
Cope, A. C., Moore, P. T. and Moore, W. R. (1959). *J. Amer. Chem. Soc.*, **81**, 3153.
Cope, A. C., Moore, P. T. and Moore, W. R. (1960). *J. Amer. Chem. Soc.*, **82**, 1744.
Cope, A. C., Ganellin, C. R., Johnson, H. W., Van Anken, J. V. and Winkler, H. J. S. (1963). *J. Amer. Chem. Soc.*, **85**, 3276.
Cope, A. C., Banholzer, K., Keller, H., Panson, B. A., Whang, J. J. and Winkler, H. J. S. (1965). *J. Amer. Chem. Soc.*, **87**, 3644; 3649.
Cram, D. J. (1965). *Fundamentals of Carbanion Chemistry*, Academic Press.
Dauben, W. G. and Berezin, G. H. (1967). *J. Amer. Chem. Soc.*, **89**, 3449.
Dunitz, J. D., Huber-Burer, E. and Venkatesan, K. (1961). *Proc. Chem. Soc.*, 463.
Dunitz, J. D. and Mugnoli, A. (1966). *Chem. Comm.*, 166.
Egmond, J. V. and Romers, C. (1969). *Tetrahedron*, **25**, 2693.
Eliel, E. L. (1962). *Stereochemistry of Carbon Compounds*, McGraw-Hill.
Eliel, E. L., Allinger, N. L., Angyal, S. J. and Morrison, G. A. (1965). *Conformational Analysis*, Wiley.
Ferguson, L. N. (1963). *The Modern Structural Theory of Organic Chemistry*, Prentice Hall.
Jackman, L. M. and Sternhell, S. (1969). *Nuclear Magnetic Resonance Spectroscopy in Organic Chemistry*, 2nd Edn., Pergamon.
Hoffmann, R. (1968). *J. Amer. Chem. Soc.*, **90**, 1475.
Kosower, E. M. and Ito, M. (1962). *Proc. Chem. Soc.*, 25.
Mislow, K. (1965). *Introduction to Stereochemistry*, Benjamin.
Mühs, M. A. and Weiss, F. T. (1962). *J. Amer. Chem. Soc.*, **84**, 4697.
Patai, S. (1964). *The Chemistry of Alkenes*, Interscience.
Roberts, J. D. and Chambers, V. C. (1951). *J. Amer. Chem. Soc.*, **73**, 5030.
Roberts, J. D., Montgomery, L. K. and Scardiglia, F. (1965). *J. Amer. Chem. Soc.*, **87**, 1917.
Sicher, J., Svoboda, M. and Zavada, J. (1964). *Tetrahedron Letters*, 15.
Wiberg, K. B., Lampman, G. M., Ciula, K. P., Connor, D. S., Shertler, P., and Lavanish, J. (1965). *Tetrahedron*, **21**, 2749.
Walsh, A. D. (1949). *Trans. Faraday Soc.*, **45**, 179.
Wittig, G. (1960). *Angewandte Chemie*, **72**, 324.

2 The conformation of rings

2.1. Cyclobutane. This as a planar structure would contain four eclipsed methylene groups. In consequence, cyclobutane takes up a non-planar structure (1), twisted by as much as *c.* 30°, which reduces torsional strain. Evidence for the non-planar structures comes from electron diffraction measurements on cyclobutane, the microwave spectrum of cyclobutyl bromide, X-ray data, and the molecular dipole moments of *cis-* and *trans*-1,3-dibromo- or 1,3-dicyano-cyclobutanes, cf. (2) and (3). Moreover, equilibration of the dibromides (2) and (3)

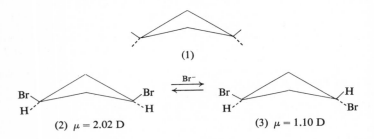

(1)

(2) $\mu = 2.02$ D   (3) $\mu = 1.10$ D

by bromide ion exchange gave a ratio: *cis/trans* = 2.07, indicating that the *cis* form (2) in which the halogen atoms are disposed equatorially, is the more stable by *c.* 2.5 kJ/mole (Wiberg and Lampman, 1966). This is, however, a relatively small energy difference, and differently substituted cyclobutanes may adopt a somewhat different conformation. From X-ray analysis, the *cis* form of cyclobutane-1,3-dicarboxylic acid (4) appears to have a puckered ring (Adman and Margulis, 1967), whilst the *trans* isomer (5) appears to be essentially planar (Margulis and Fisher, 1967). The X-ray data also indicate some carbon–carbon bond lengthening in cyclobutane derivatives; values as high as 156.7 pm

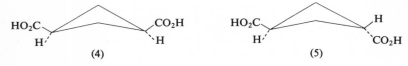

(4)   (5)

are reported in comparison with 154 pm for the C–C bonds of cyclohexane.

2.2. Cyclopentane. From combustion data this has a strain energy of
c. 27.3 kJ/mole, and the torsional strain due to five eclipsed methylene
units in a planar structure must evidently be reduced by twisting of the
ring (cf. Brutcher *et al.*, 1959). The microwave spectrum and thermodynamic properties indicate a puckered ring for cyclopentane and
X-ray data for some natural products containing a cyclopentane unit
indicate a bond angle of *c.* 105°, i.e. smaller than tetrahedral. Heats of
combustion indicate that *cis*-1,3-dimethylcyclopentane (6) is more
stable than the *trans* form by *c.* 2.1 kJ/mole which is accounted for in

(6)

(7)

the 'envelope' conformation (7) which by bending the ring minimises
Me:H and H:H interactions.

(8)

(9)

(10)

In cyclopentane itself the effect of twisting a particular –CH$_2$– element
is transmitted around the ring in a sequence amounting to a process of
pseudo-rotation. Introduction of a trigonal centre as in cyclopentanone
removes a number of eclipsing bonds, and a half-chair (8) rather than the
envelope conformation (7) becomes the preferred form.

2.3. The cyclohexane ring. This is able to adopt the chair conformation
(9) (Hassel, 1953) in which the substituents are staggered as shown in
(10). Exact measurements indicate a slight flattening of the perfect chair,
and widening of the ring angles to 111.5°, i.e. slightly greater than tetrahedral. The cyclohexane chair structure was derived from electron

diffraction data, but the alternative boat conformation which is flexible, i.e. of higher entropy, is also excluded by the thermodynamic properties of cyclohexane.

The chair cyclohexane conformation (9) distinguishes axial (ax.) and equatorial (eq.) orientations for substituents. This different steric environment results in some difference in reactivity of axial and equatorial substituent groups which provided the basis for the development of the subject of conformational analysis by Barton (1950, 1953).

2.4. Conformational inversion. The cyclohexane ring may, however, undergo conformational inversion, and chair–chair inversion of a cyclohexane (11; R = H or other substituent) may be depicted:

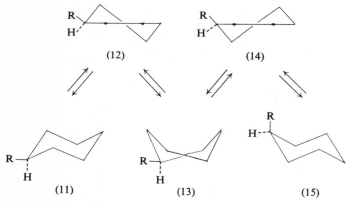

or, in energy terms:

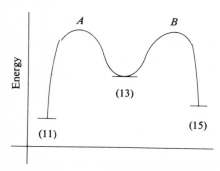

where *A* and *B* correspond to the flattened ring states (12, 14) in the process of inversion, and (13) is the twisted boat form.

Below −100°C the n.m.r. spectrum of undecadeuterocyclohexane develops two separate signals corresponding to the proton in an equatorial (16) or axial (17) environment. At higher temperatures inversion: (16)⇌(17) results in an averaged signal being observed (Anet and Bourn, 1967).

(16)   H$_{(eq.)}$, τ 8.4

(17)   H$_{(ax.)}$, τ 8.88

From the temperature change of line widths the free energy and enthalpy of activation for chair–chair inversion are derived as 43.3 and 45.8 kJ/mole respectively. This is a measure of the energy required to pass through the flattened structure (12), or (14).

2.5. Conformational equilibria.

The order of magnitude of the activation energy for chair–chair inversion of cyclohexane derivatives permits rapid interconversion at the ordinary temperature between the alternative chair conformations, (18) and (19), of a substituted cyclohexane. The position of the equilibrium between (18) and (19) depends on the free energy difference between these conformers in which the substituent R occupies respectively an equatorial and an axial orientation. The equilibrium between conformations (18) and (19), which reflects the effective steric requirements of the substituent R, has been examined for a number of substituents in various ways, cf. Eliel (1965), Franklin and Feldkamp, (1965).

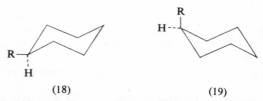

(18)   (19)

In many cases, e.g. where R = OH, OR, halogen, the methine proton on the carbon atom carrying the R-group is sufficiently deshielded to be recognised in the n.m.r. spectrum. Conformations (18) and (19) may then, in principle, be distinguished by the generally smaller chemical shift for an axial ($\delta_{ax.}$) than for an equatorial ($\delta_{eq.}$) methine proton. At the ordinary temperature, however, rapid inversion between (18) and (19) results in an averaged spectrum showing an averaged proton chemical shift, $\delta_{av.} = \delta_{ax.} \cdot N_{ax.} + \delta_{eq.} \cdot N_{eq.}$, where $N_{ax.}$ and $N_{eq.}$ are the mole fractions of (18) and (19) respectively. To derive $N_{ax.}$ and $N_{eq.}$, $\delta_{ax.}$ and $\delta_{eq.}$ may be obtained by cooling so as to reduce the rate of inversion, but more often values for $\delta_{ax.}$ and $\delta_{eq.}$ have been taken from reference substances of fixed conformation. For this purpose the *trans-* (20),

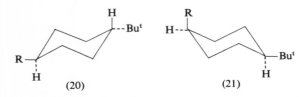

(20)   (21)

and *cis*-t-butyl derivative (21), have frequently been used. The large steric requirements of the t-butyl group maintain this substituent in an equatorial orientation, and, assuming no ring deformation, the R-group will be oriented equatorially or axially as in (20) and (21), and the methine proton axially and equatorially respectively.

In this way the position of the equilibrium: (18)⇌(19) and the free energy difference ΔG for a variety of substituents has been derived, cf. Table 2.1. The values indicate the relative steric requirements of substituents attached to a cyclohexane ring.

TABLE 2.1 *Conformational free energy differences* (kJ/mole)

R	ΔG	R	ΔG	R	ΔG
Me	7.15	OAc	2.0	CN	0.8
i-Pr	9.3	CO$_2$Me	4.6	Cl	2.1
t-Bu	20				

However, despite its relative remoteness the 4-t-butyl group is not without influence on the chemical shift of the methine proton being examined, and the use of the t-butyl substituent as a conformational holding group in n.m.r. studies is therefore not without objections

(Eliel and Martin, 1968). A preferred alternative is to measure relative peak areas for the axial and equatorial methine proton at a low temperature where conformational inversion is slow on the n.m.r. time scale (Jenner *et al.*, 1969). A different method (Eliel and Reise, 1968) based on equilibration:

and gas chromatogram (g.c.) analysis may be employed in cases where the substituent R assists removal of the methine proton, e.g. R = CO_2H or COCl with acid catalysed equilibration, or R = CO_2Me or COMe with equilibration by alkaline catalysis. The free energy difference between the conformations is then derived from the equilibrium constant.

cis- [22], and *trans*-1,3-di-t-butyl cyclohexane, [23], may be equilibrated by heating with a palladium catalyst, i.e. by a process of catalytic

dehydrogenation/re-hydrogenation. To avoid an axially oriented t-butyl group the 1,3-*trans* isomer must necessarily take up the boat conformation (23), and from the change in ratio (22)/(23) with temperature, the difference in free energy and enthalpy between the boat and chair cyclohexane conformation may be derived. From this experiment and from heats of combustion and other data the following best values are obtained: chair→boat, $\Delta H + 26$ kJ/mole, $\Delta S + 12.2$ J/deg mole (Allinger and Freiberg, 1960).

The chair→boat inversion is characterised by a large positive entropy change since the chair is rigid and the twisted boat is a highly flexible structure.

2.6. Cyclohexene.   X-ray data show that the cyclohexene ring preferentially adopts a half-chair conformation, (24). At −170°C the n.m.r. spectrum of the deuterocyclohexene: (25)⇌(26) is resolved to show distinct signals for the protons H* which through rapid inversion between conformations (25) and (26) are seen at higher temperatures as an averaged signal at *c.* τ8.4 (Anet and Huq, 1965). The enthalpy of activation for half chair (25)→half chair (26) inversion derived from the n.m.r. data is 22.2 kJ/mole, i.e. cyclohexene inverts much more rapidly than cyclohexane.

(24)

(25)   (26)

From heats of combustion and palladium catalysed equilibration relative energy data are available also for a range of bicyclic structures, cf. Table 2.2 (Eliel *et al.*, 1965).

2.7. The decalins, hydrindanes and other bicyclic structures.   *Trans* and *cis* fusion of chair cyclohexane units in the decalins (27 and 28; R = H) involves no change in dihedral angle, cf. (27*a*) for *trans* and (28*a*) for *cis* fusion. The relatively higher enthalpy of *cis*-decalin (28) is due to more severe non-bonded interactions in the *cis* than in the *trans* form. Projection along any C–C bond of cyclohexane in the chair conformation reveals a four-carbon unit oriented as in the gauche conformation of butane. A gauche oriented butane (29) represents a potential energy of 3.35 kJ/mole above the *anti*-conformation (30). Inspection of *cis*-decalin (28) shows the presence of three more gauche butane type interactions than in *trans*-decalin (27). This represents an energy of 3 × 3.35 = 10.1 kJ/mole above that of *trans*-decalin in good agreement with the experimental value of 11.5 kJ/mole (cf. Table 2.2).

In *trans*- (27) or *cis*-decalin (28), an R-substituent larger than hydrogen introduces increased non-bonded interactions. These, however, are more serious in the *trans* isomer where the R-group is axial to both rings.

TABLE 2.2 *Enthalpy differences for cis- and trans- forms of various fused ring structures*

	kJ/mole
trans-→cis-decalin, (27)→(28), R = H	11.5
trans-→cis-10-methyldecalin, (27)→(28) R = CH$_3$	5.9
trans-→cis-10-methyl-1-decalone, (31)→(32)	0.8
cis-→trans-hydrindane, (34)→(33)	4.2
cis-→trans-perhydroazulene, (37)→(36)	small
cis-→trans-bicyclo[3,3,0]octane, (39)→(38)	29.6

Thus the R-substituent raises the energy of the *trans*-decalin more than that of the *cis* form and narrows the energy gap from 11.5 kJ/mole when R = H to 5.9 kJ/mole when R = CH$_3$. A trigonal centre, such as a carbonyl group, on the other hand, reduces axial interactions which are more serious in the *cis*-decalin series. In consequence, the energy difference between *trans*- and *cis*-10-methyldecal-1-one (31) and (32), becomes very small indeed.

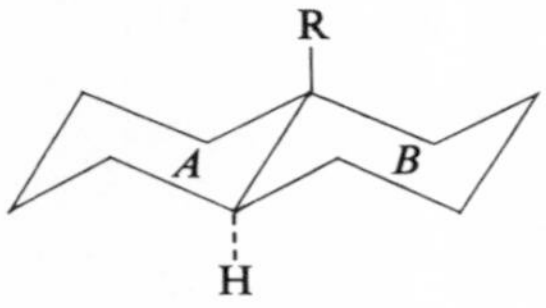

(27)

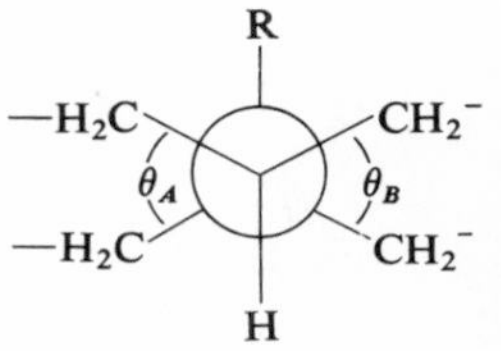

(27a)

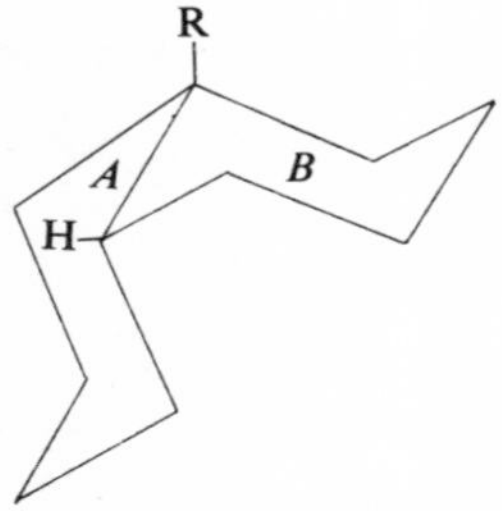

(28)

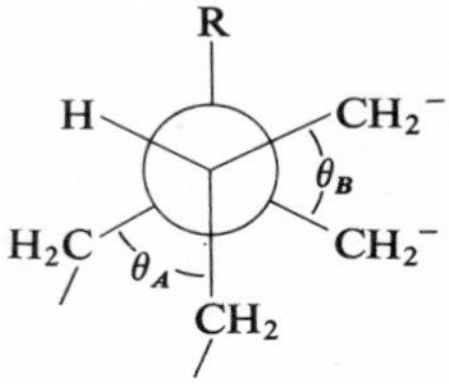

(28a)

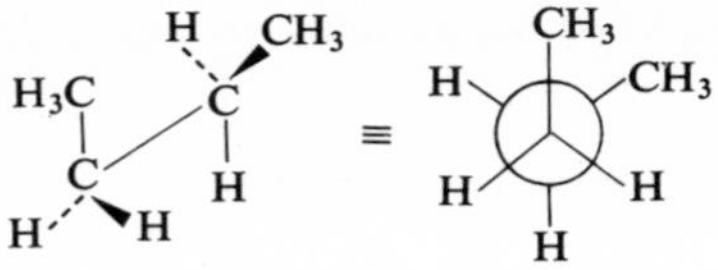

(29)

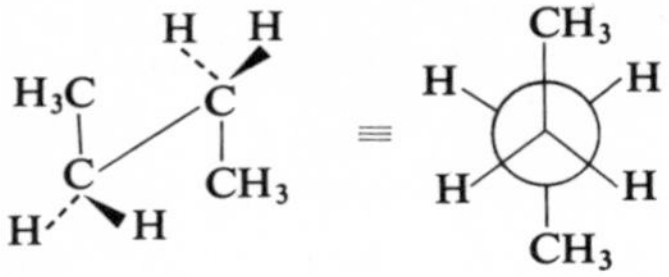

(30)

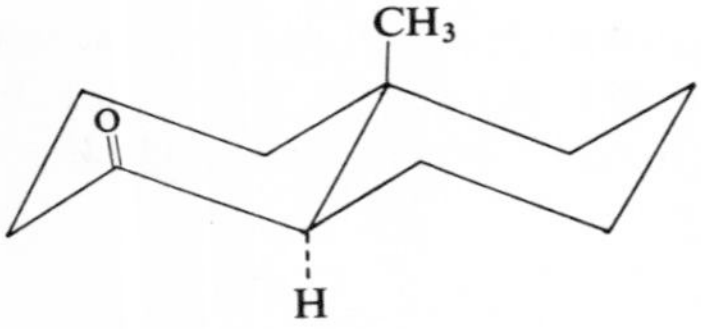

(31)

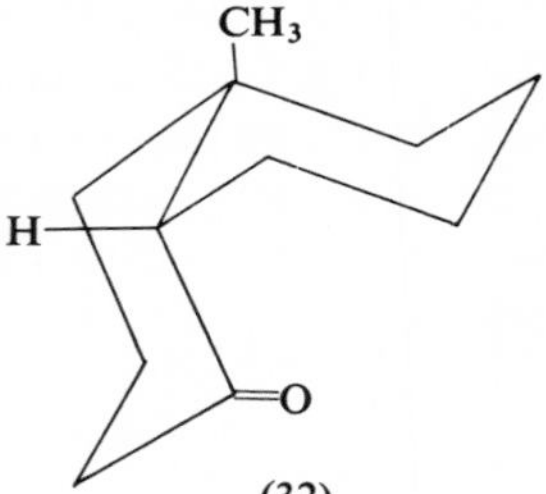

(32)

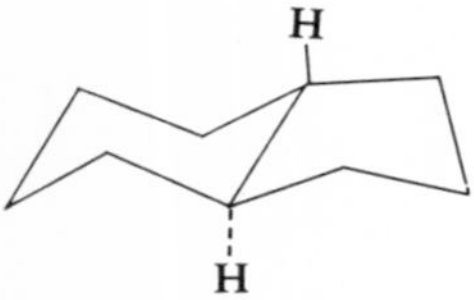

(33)

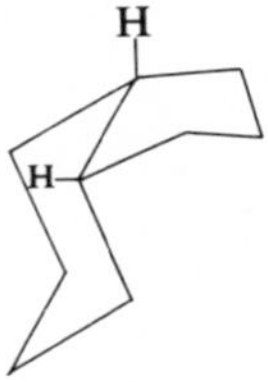

(34)

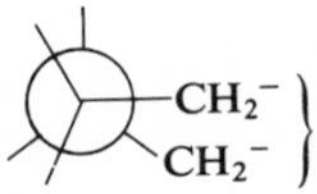

(35)

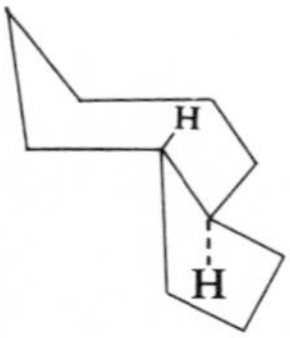

(36)

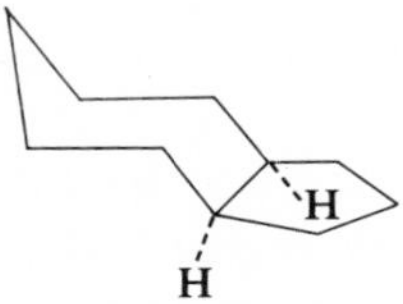

(37)

The near planarity of the cyclopentane ring means that a fusion of a cyclohexane and a cyclopentane ring as in *trans*- (33) and *cis*-hydrindane (34) introduces a bridgehead unit (35), and necessarily some change in the normal dihedral angle in both rings. Inspection of (27a) and (28a) shows that in (27a) compression of θ_B widens θ_A, whilst in (28a) θ_A and θ_B alter in the same sense. For this reason, *cis*-hydrindane, which corresponds to the situation in (28a) introduces less torsional strain, and is therefore more stable than *trans*-hydrindane.

Larger rings accommodate a near-eclipsed bridgehead unit more easily, and in consequence *cis*- and *trans*-perhydroazulene, (36) and (37), are of almost equal stability.

Trans fusion of two five-membered rings, on the other hand, involves very considerable torsional strain. This is illustrated by the bicyclo-[3,3,0]octanes, (38) and (39), and the data of Table 2.2.

cis-Decalin is a relatively flexible structure, and a substituted *cis*-decalin, e.g. (40), may adopt two conformations, (41) and (42), described respectively as the steroid, and non-steroid conformation, since the former conformation occurs in coprostane.

(38)   (39)

(40)   (41)   (42)

Rapid interconversion between alternative conformations of this kind, e.g. (44) and (45), is apparent from the ^{19}F n.m.r. spectrum of 2,2-difluoro-*cis*-decalin (43); the two fluorine atoms are distinguished as an *AB*-system only below −78°C (Roberts, 1966). Similarly, whilst axial

and equatorial groups of protons are seen in the n.m.r. spectrum of the
rigid *trans*-decalin, the low temperature spectrum of *cis*-decalin is not
resolved in this way.

2.8. Cyclohexane structures containing a boat conformation. Although
in a majority of cyclohexane derivatives the chair is the preferred conformation, this situation may be reversed, e.g. as in (23) by the presence
of a very large axial substituent, or by the presence of one or more trigonal centres. Cyclohexan-1,4-dione is a long-known instance of the
latter; X-ray analysis shows, in the solid, a preferred conformation
(46) with an angle of 154° between the two $\diagdown$C=O bonds, and a net
molecular dipole moment of *c*. 1 Debye unit.

(43)

(44)   (45)

(46)

Examples of boat-type conformations are found amongst derivatives
of both *cis*- and *trans*-decalones, e.g. in the steroid ketone (47), despite
the dipole repulsion between the –C=O and C—Br bonds, or in (48)
where there is more serious interaction between the C-4—Me and
C-6—OH groups in the alternative chair–chair form. In both (47) and
(48) the presence of a trigonal group at C-3 reduces the energy of chair→
boat inversion.

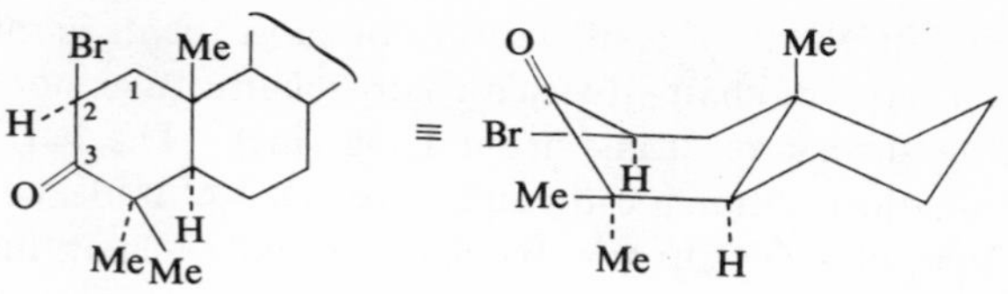

(47)

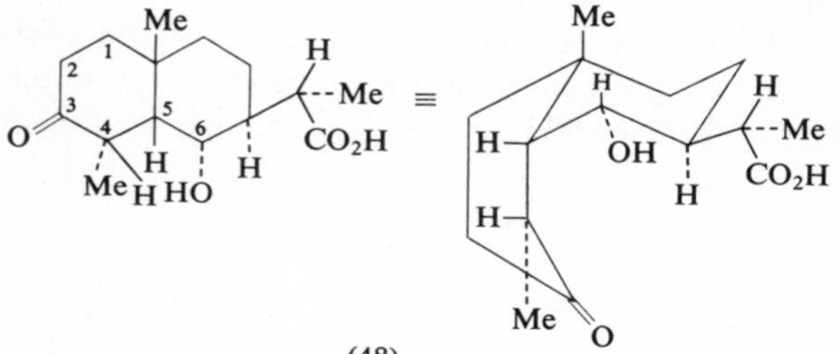

(48)

2.9. Cycloheptane. This has no rigid and obviously preferred conformation corresponding to the cyclohexane chair conformation. In the chair-type conformation (49) there is an eclipsed ethane unit at 4,5, and a rather close approach (128) pm of the hydrogen marked (*) at 3 and and 6. However, the ring is flexible, and a twist chair form (50) is possible which reduces the torsion at 4,5 and increases the separation of the 3,6-hydrogens to 186 pm. Down to −170°C the two fluorine substituents in 1,1-difluorocycloheptane (51) shows no chemical shift difference, indicating very rapid conformational change (Roberts, 1966). In

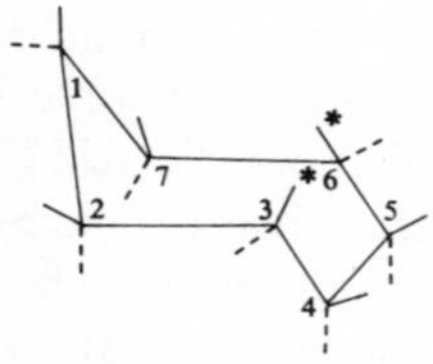

(49)

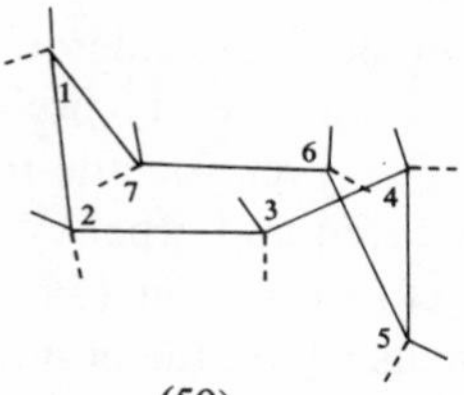

(50)

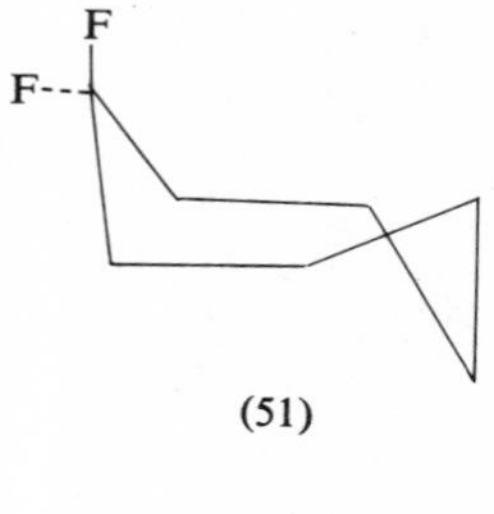

(51)

cycloheptane, inversion of configuration of substituents is rather easily possible by a series of chair→twisted chair→chair conversions achieved by displacing successive atoms around the rings. The barrier to this process of pseudo-rotation is estimated to be *c*. 8.4 kJ/mole, i.e. considerably less than the 46 kJ/mole barrier for chair→chair inversion in cyclohexane.

In cycloheptene the olefinic bond introduces an element of rigidity, and the available evidence points to a preferred chair conformation. In the diacetate (53) the benzyl protons appear in the n.m.r. spectrum as a doublet at $\tau 4.05$, $J = 9$ Hz., i.e. with coupling indicating one adjacent

(52)   (53)

proton at a small dihedral angle and one at a dihedral angle of about 90°. Also, there is evidence of a conformational preference in the low temperature n.m.r. spectrum of (52) which shows splitting of the signal for the CH_2^* protons at *c*. $\tau 8.75$.

2.10. Cyclo-octane. Considerations of strain and bond interactions do not indicate any clearly preferred conformation for cyclo-octane. X-ray analyses on cyclo-octane-*trans*-1,2,dicarboxylic acid (55) (Dunitz and Mugnoli, 1966) and other cyclo-octane derivatives have established a chair-boat conformation (54) in these substances. The chair-boat conformation also gives the best interpretation of the n.m.r. spectrum of

(54)   (55)

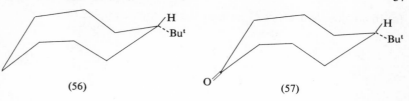

(56)   (57)

deuterocyclo-octanes. However, the ring is flexible and both t-Bu-cyclo-octane (56) and 5-t-Bu-cyclo-octanone (57) show two (unequal) n.m.r. bands due to the t-butyl group at low temperatures (Anet and Jacques, 1966), i.e. there is evidence of more than one conformation.

2.11. Conformation and reaction. The influence of conformation on reaction is seen in two principal ways.

(i) Steric environment may contribute hindrance or acceleration to the reactivity of a substituent according as formation of the transition state of reaction involves an increase or a decrease in bulk at the reaction site.

(ii) By maintaining a definite angular relation between adjacent groups, conformation will determine the probability of reactions with a definite steric requirement such as an *anti*-coplanar or *syn*-coplanar relation between the groups concerned in reaction.

These principles are well illustrated in the chemistry of cyclohexanes where the chair conformation (58) distinguishes axial (ax.) and equatorial (eq.) groups:

(ax.)   (ax.)

(eq.)--   --(eq.)

(eq.)   (eq.)

(ax.)

(ax.)

(eq.)--   (eq.)   --(eq.)

(ax.)   (ax.)

(58)

The influence of the steric requirements of the transition state is illustrated in the following examples.

(*a*) In alkaline hydrolysis of an ester the reaction complex (59):

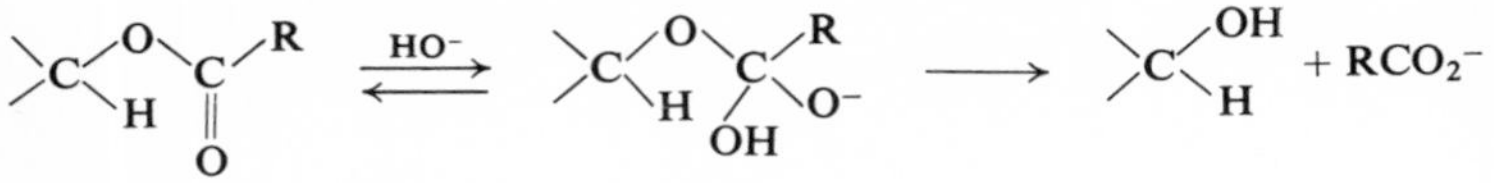

(59)

is formed with an increase in bulk.

(*b*) In the reaction complex for acetolysis of a toluene-*p*-sulphonate, (60), on the other hand, two groups (OTs and OAc) are relatively remotely attached, and the carbon centre is effectively trigonal, i.e. there is a decrease in compression.

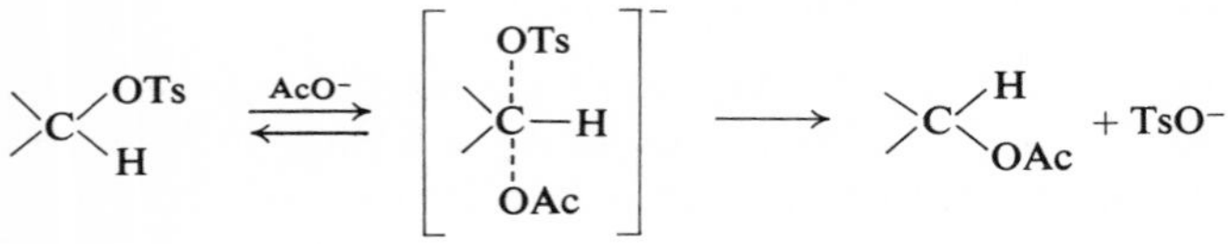

(60)

(*c*) The rate determining step in chromic acid oxidation of a secondary alcohol is proton removal as in (61):

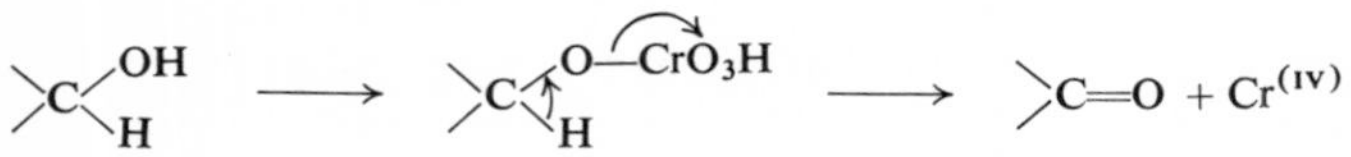

(61)

in which the co-ordination at the reaction site passes from tetrahedral to trigonal, i.e. with a decrease in bulk.

2.12. Rates of hydrolysis, acetolysis and oxidation. Less compressed, i.e. equatorial, substituents react more rapidly where there is an increase in bulk at the transition state as in ester hydrolysis. On the other hand, acetolysis of a sulphonate, or oxidation of an alcohol, exhibit steric acceleration when the displaced group (OTs or OH) is in a more compressed, i.e. an axial environment.

The orders of magnitudes of these effects, which are not large, are indicated in Tables 2.3 and 2.4, cf. Eliel *et al.*, (1965).

TABLE 2.3   *Relative rates of hydrolysis of decalyl esters and of acetolysis of decalyl toluene-p-sulphonates†*

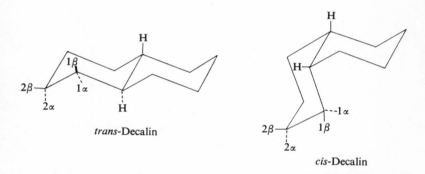

trans-Decalin   *cis*-Decalin

(i) Relative rates of alkaline hydrolysis of acid succinates:

trans-decalyl-			*cis*-decalyl-		
2α	ax.	0.13	2α	eq.	0.95
2β	eq.	1.0	2β	ax.	0.58

(ii) Relative rates of acetolysis of toluene-*p*-sulphonates:

trans-decalyl-			*cis*-decalyl-		
1α	ax.	14.6	1α	ax.	2.0
1β	eq.	0.28	1β	eq.	1.1
2α	eq.	0.53	2α	eq.	2.0
2β	ax.	1.65	2β	ax.	2.4

The differentiation in reactivity of axial and equatorial groups is less marked in the flexible *cis*-decalin derivatives than in the rigid *trans*-decalin series.

† The indication of steric orientation of groups by the prefix α or β is discussed in §1.2.

TABLE 2.4 *Relative rates of chromic acid oxidation of axial and equatorial cholestanols*

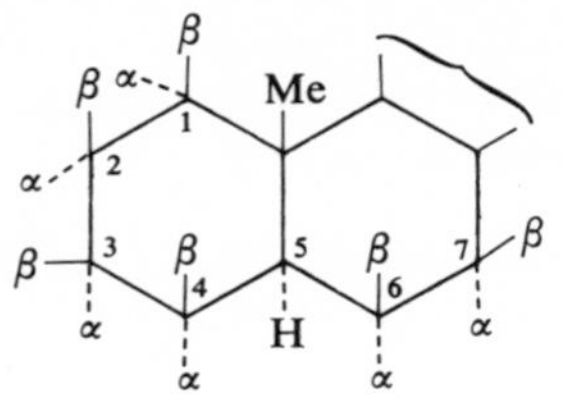

Equatorial OH		Axial OH	
1β	9.7 †	1α	13†
2α	1.3	2β	20
3β	1.0	3α	3.0
4α	2.0	4β	35
6α	2.0	6β	36
7β	3.3	7α	12.3

† rate relative to 3β OH as standard.

2.13. Elimination reactions. Axially oriented groups are also more readily lost by elimination, e.g. in the deamination:

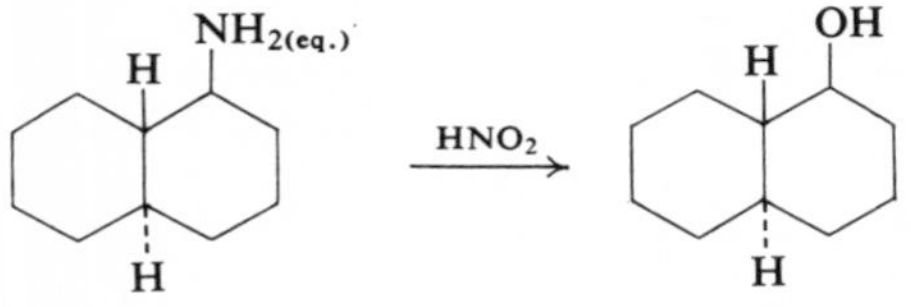

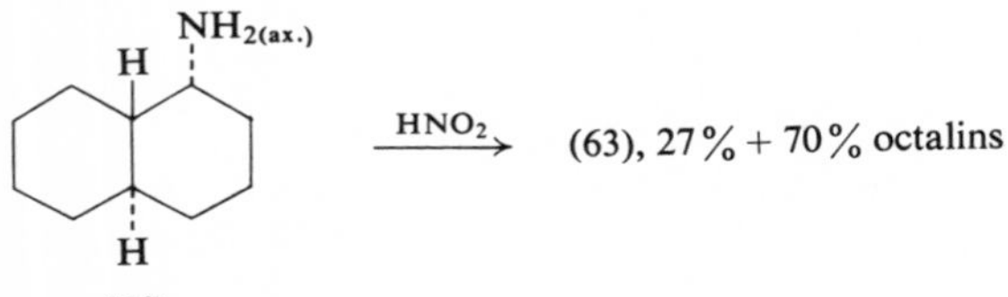

where the axial amine (64) gives the octalin in high yield whereas the equatorial amine (62) undergoes displacement to give the decalol (63). The stereochemistry of the decalol product (63) formed arises from reaction of the carbonium ion intermediate in deamination from the less hindered side, i.e. equatorially. The predominant elimination reaction observed with the axial amine (64) is due to the *anti* and coplanar relationship between the vacant orbital left by the departing N_2 residue and an adjacent axial hydrogen, i.e. favourable for olefin formation.

This type of conformational influence is most marked in concerted 1,2-elimination reactions. The process of debromination of vicinal dibromides by means of iodide ion:

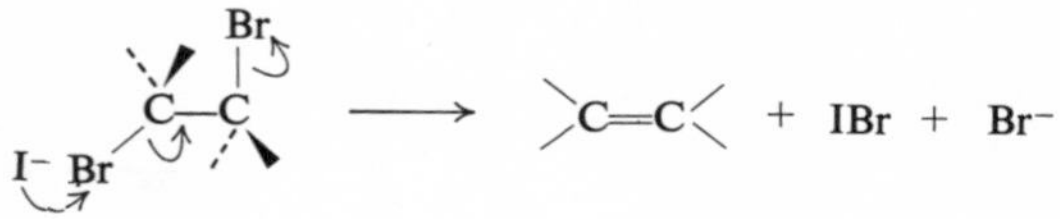

provides a particularly clear example. Thus the 5α, 6β-dibromide (65) in which both bromo-substituents are axial and in an *anti* coplanar relationship, is rapidly debrominated to (66) (Barton and Rosenfelder, 1951; Alt and Barton, 1954). The isomeric 5β, 6α-dibromide (67),

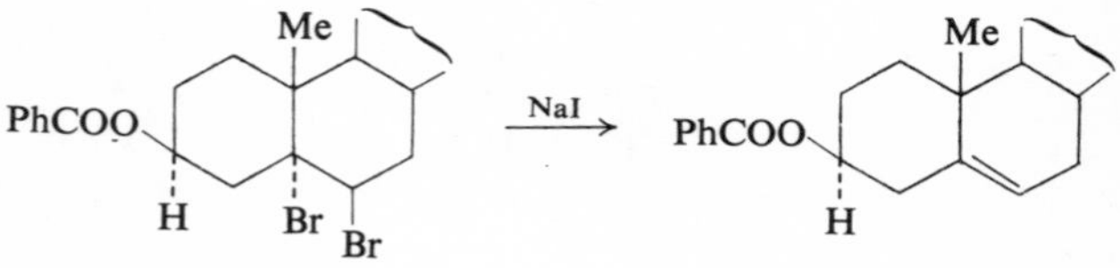

(65)   (66)

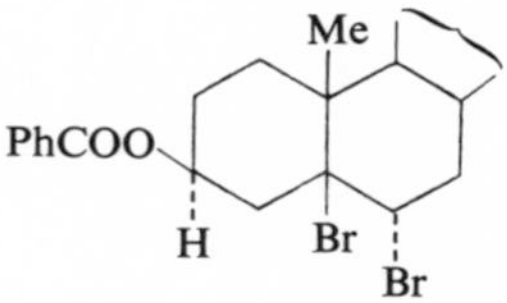

(67)

in which the bromo-substituents subtend with each other an angle of ~60°, on the other hand, fails to react in this way with iodide ion.

The operation of this type of stereoelectronic control must, however, clearly depend upon the requirements of the particular reaction mechanism. Thus by contrast with the examples (65) and (67) it has been found that the isomeric bromohydrins (68), (69) and (70) react with zinc with

equal ease to give the olefinic product (71) (James *et al.*, 1955). In this
case the elimination:

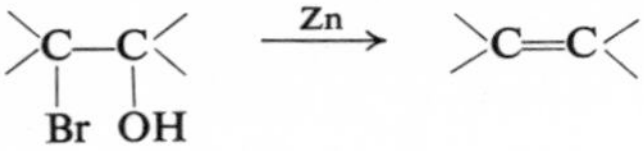

$$\overset{|}{\underset{Br}{\overset{|}{C}}}-\overset{|}{\underset{OH}{\overset{|}{C}}}\xrightarrow{\ Zn\ }\ \overset{|}{C}=\overset{|}{C}$$

which presumably may involve an alkyl zinc intermediate, does not appear
to have exacting steric requirements.

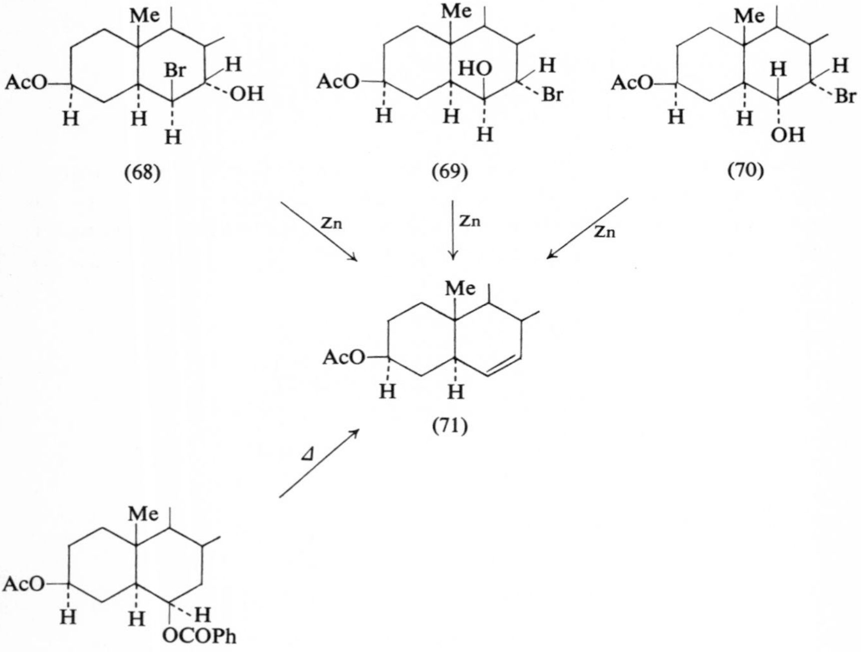

(68)   (69)   (70)

Zn   Zn   Zn

(71)

Δ

(72)

The thermal elimination of benzoic acid from (72) to give (71) is
an example of a group of reactions in which the eliminated groups are
necessarily *syn* related in the reaction transition state:

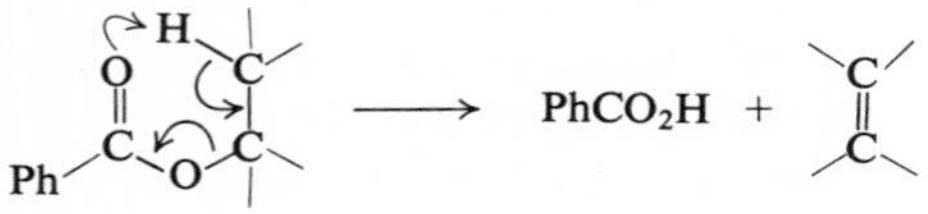

The Cope elimination:

is a further example of the same kind. This and the related Hofmann elimination are more fully discussed together in chap. 3.

References

Adman, E. and Margulis, T. N. (1967). *Chem. Comm.*, 641.
Allinger, N. L. and Freiberg, M. (1960). *J. Amer. Chem. Soc.*, **82**, 2393.
Anet, F. A. L. and Huq, M. L. (1965). *J. Amer. Chem. Soc.*, **87**, 3147.
Anet, F. A. L. and Jacques, M. St (1966). *J. Amer. Chem. Soc.*, **88**, 2586.
Anet, F. A. L. and Bourn, A. J. R. (1967). *J. Amer. Chem. Soc.*, **89**, 760.
Barton, D. H. R. and Rosenfelder, W. J. (1951). *J. Chem. Soc.*, 1048, cf. Alt, G. H. and Barton, D. H. R. (1954). *J. Chem. Soc.*, 4284.
Barton, D. H. R. (1953). *J. Chem. Soc.*, 1027; cf. *Experientia* (1950), **6**, 316; (1955), **11**, 121.
Brutcher, F. V., Roberts, T., Barr, S. J. and Pearson, N. (1959). *J. Amer. Chem. Soc.*, **81**, 4915.
Dunitz, J. D. and Mugnoli, A. (1966). *Chem. Comm.*, 166.
Eliel, E. L. (1965). *Angewandte Chemie* (Internat. Edn.), **4**, 761.
Eliel, E. L., Allinger, N. L., Angyal, S. J. and Morrison, G. A. (1965). *Conformational Analysis*, Wiley.
Eliel, E. L. and Reise, M. C. (1968). *J. Amer. Chem. Soc.*, **90**, 1560.
Eliel, E. L. and Martin, R. J. L. (1968). *J. Amer. Chem. Soc.*, **90**, 689.
Franklin, N. G. and Feldkamp, H. (1965). *Angewandte Chemie* (Internat. Edn.), **4**, 774.
Hassel, O. (1953). *Quarterly Reviews*, **7**, 221.
James, D. R., Rees, R. W. and Shoppee, C. W. (1955). *J. Chem. Soc.*, 1370.
Jenner, F. R., Bushweiler, C. H. and Beck, B. H. (1969). *J. Amer. Chem. Soc.*, **91**, 344.
Margulis, T. N. and Fisher, M. (1967). *J. Amer. Chem. Soc.*, **89**, 223; cf. Margulis, T. N. (1969). *Chem. Comm.*, 215.
Roberts, J. D. (1966). *Chemistry in Britain*, 529.
Wiberg, K. D. and Lampman, G. M. (1966). *J. Amer. Chem. Soc.*, **88**, 4429.

3 The influence of ring size on the reactivity of substituent groups

3.1. Reactions influenced by ring size. The influence of ring size on the reactivity of cycloalkyl derivatives is illustrated in a number of ways.

(i) The rate of displacement reactions of the type:

$$(CH_2)_{n-1} \underbrace{}CHX \xrightarrow{\;Y^-\;} \left\{ (CH_2)_{n-1} \underbrace{} \overset{Y}{\underset{X}{C}}{-}H \right\}^{-} \longrightarrow (CH_2)_{n-1}\underbrace{}CHY$$

varies with the ring size.

(ii) Ring size influences both the rates of addition to cycloalkanones, and the position of the equilibrium:

$$(CH_2)_{n-1}\underbrace{}C{=}O \underset{\longleftarrow}{\overset{HY}{\longrightarrow}} (CH_2)_{n-1}\underbrace{}C\overset{OH}{\underset{Y}{\diagdown}}$$

e.g. $HY = HCN$

(iii) A marked influence of ring strain is seen in the rates of addition of various reagents to cycloalkenes.

(iv) The rate of enolisation of cycloalkanones is a function of ring size.

(v) The phenomenon of transannular displacement reactions accompanied by hydrogen transfer across the rings:

$$H{-}C{-}H \underbrace{\overset{(CH_2)_x}{\underset{(CH_2)_y}{}}}CH{-}X \xrightarrow{\;HY\;} H{-}Y{\cdots}\overset{(CH_2)_x}{\underset{\underset{(CH_2)_y}{H}}{C}}{\cdots}H{\cdots}CH{\cdots}X \longrightarrow$$

$$Y{-}CH\underbrace{\overset{(CH_2)_x}{\underset{(CH_2)_y}{}}}HCH + HX$$

is observed mainly in medium-sized rings ($n = 8$ to 11) where hydrogen substituents are turned inside the ring, and so are brought near to the cationic site which is developed by the departure of the 'leaving group'.

(vi) In vicinally substituted rings (1) the angular relation of adjacent groups, X and Y, which are concerned in an elimination reaction, or in cyclisation, or in physical interaction, is a function of ring size.

$$(CH_2)_n \begin{array}{c} CH{-}X \\ \\ CH{-}Y \end{array}$$

(1)   (2)   (3)

Examples include the case where $X = \overset{+}{N}R_3$, $Y = H$; or $X = -\overset{\overset{\bar{O}}{+|}}{N}R_2$, $Y = H$ as in the Hofmann and Cope elimination reactions, or where e.g. $X = -O{-}SO_2R$ and an adjacent group, $Y = NHCOR$ assists displacement *via* an oxazolidine intermediate, e.g. (2)→(3).

3.2. Displacement reactions. Table 3.1 summarises the relative rates of acetolysis of the cycloalkyl toluene-*p*-sulphonates, $(CH_2)_{n-1}CHOTs$.

The reaction rates show two maxima, one between $n = 4$ and 5, and a second between $n = 7$ and 11. Beyond $n = 14$ the rate is of the same order as for a typical acyclic secondary toluene-*p*-sulphonate, e.g. $(n\text{-}C_3H_7)_2$-CHOTs, cf. Brown and Ham (1956); Eliel (1962); Sicher (1962).

TABLE 3.1   *Rates of acetolysis of cycloalkyl toluene-p-sulphonates at 25°C relative to cyclohexyl toluene-p-sulphonate, and ν_{CO} for cycloalkanones*

n	rel. rate	ν_{CO} (cm^{-1})	n	rel. rate	ν_{CO} (cm^{-1})
3	10^{-5}	1815	11	48.9	1709
4	11.3	1791	12	3.25	1713
5	14.0	1748	13	3.5	1713
6	1.0	1716	14	1.3	1714
7	25.3	1705	15	2.19	1715
8	191	1703	17	2.17	—
9	172	1703	20	1.80	—
10	380	1704	$(n\text{-}C_3H_7)_2$CHOTs	1.33	1718

The extremely small rate of acetolysis of cyclopropyl toluene-*p*-sulphonate is to be expected from the relatively high s-character, and hence the strength of the C_3H_5–OTs bond (cf. chap. 1). The rate for the cyclohexyl toluene-*p*-sulphonate is very close to the rate for an acyclic alkyl derivative, which reflects the normal carbon–carbon bond angle, and the absence of bond opposition strains in the cyclohexane ring. The considerable rate enhancement in the cases $n = 4$ and 5, and $n = 7$ to 11 is due to steric assistance to the acetolysis:

which arises from three main causes.

As the $\rangle$CH–OTs bond is stretched in solvolysis there is (i) a reduction in compression because of separation of the groups, and (ii) a change from sp^3 towards sp^2 geometry at the reaction site. The accompanying change in torsional strain and non-bonded interactions in some cases provides steric assistance to the displacement. The operation of these factors for displacement in cyclopentyl toluene-*p*-sulphonate (4)→(5), in comparison with cyclohexyl toluene-*p*-sulphonate: (6)→(7) is illustrated:

(4)   (4a)   (5)   (5a)

(6)   (6a)   (7)   (7a)

Comparison of (4a) and (5a) shows that formation of the transition state in the cyclopentyl case (5) is accompanied by reduction in H–H interactions as the reaction site acquires trigonal geometry. In the cyclohexyl derivatives (6) the projections (6a) and (7a) show that in the intermediate (7) there is eclipsing of a pair of hydrogen substituents.

Thus the displacement (4)→(5) is accompanied by a reduction in compression which is not the case for (6)→(7). This difference lies behind the relatively greater rate of acetolysis of the cyclopentyl toluene-*p*-sulphonate.

A second factor which may assist the displacement arises in the medium-sized rings where the carbon–carbon bond angles are already widened somewhat in reducing torsional strain. This establishes in the parent toluene-*p*-sulphonate a carbon–carbon bond angle, and state of hybridisation, relatively nearer to that corresponding with the trigonal geometry of the transition state for displacement. Thus in the range of ring sizes: $n = 7$ to 11, acetolysis of a toluene-*p*-sulphonate is assisted for this reason.

This bond angle hybridisation factor is reflected also in the carbonyl stretching frequency of the cycloalkanones (cf. Table 3.1). In fact log (rate of acetolysis) and ν_{CO} for the cycloalkanones shows a reasonably linear inverse relation over the range $n = 6$ to 15. However, although this parallel is of interest in drawing attention to the importance of the change in bond angle, the values of ν_{CO} cannot reflect the changes in torsional strain and non-bonded interactions which also contribute to the energy of reaching the transition state for displacement (cf. Foote, 1964; Schleyer, 1964).

3.3. The cyclobutyl- and cyclopropylmethyl cation. In the case of cyclobutyl toluene-*p*-sulphonate acetolysis is assisted by a factor of a different kind, namely by the possibility of charge delocalisation in the transition state and reaction *via* a non-classical type of carbonium ion. Cyclobutyl toluene-*p*-sulphonate, and other cyclobutyl derivatives are found to react with rearrangement *via* a reaction intermediate which leads to products containing the cyclopropyl methyl (9), and but-3-enyl (10) carbon skeleton as well as to the expected cyclobutyl derivative. Moreover, in solvolyses, cyclopropyl methyl and cyclobutyl derivatives give the same mixture of products. Cyclobutyl- (8, Y = Cl), and cyclopropylmethyl- (9, Y = Cl) chlorides readily rearrange to the same mixture of chlorides (8), (9), and (10), Y = Cl. Acetolysis of cyclobutyl- (8, Y = OTs) and cyclopropyl methyl (9, Y = OTs) toluene-*p*-sulphonates gives the same mixture of acetates, viz.: (8), 22%; (9), 65%; and (10), 13%; Y = OAc. Deamination of the corresponding amines (8) and (9), Y = NH_2, gives a similar set of alcohols. Further, in displacements an isotopic label as in (9, * = CD_2 or $^{14}CH_2$) becomes distributed over all four carbon atoms in the products (Roberts, 1951; 1959).

Cyclobutyl bromide solvolyses $(k_1, 0.015 \text{ h}^{-1})$ at almost the same rate as the reactive allyl bromide $(k_1\ 0.013 \text{ h}^{-1})$, and for cyclopropyl methyl bromide solvolysis is appreciably faster, $(k_1, 0.34 \text{ h}^{-1})$.

(8)   (9)   (10)

Evidence bearing on the cyclopropyl methyl cation which is the intermediate in these reactions is obtained from considering the n.m.r. spectrum of the carbonium ion (12) formed from (11) in $FSO_3H/SO_2/SbF_5$ solution (Olah *et al.*, 1965). In (12) the two methyl groups are found to be non-equivalent in the n.m.r. spectrum indicating that (12) represents

(11)   (12)   (13)

the preferred conformation. This conformation, in which the vacant orbital represented by the cationic charge is parallel to the plane of the cyclopropane ring, permits overlap between this vacant orbital and the electrons of the C–C bonds of the ring with consequent charge delocalisation and stabilisation of the conformation (13). The charge delocalisation is seen in the ^{13}C n.m.r. chemical shift of the cationic carbon $-^{+}CMe_2$ $(\delta = -86.8 \text{ p.p.m. from } CS_2)$ in (13) which is a good deal smaller than is found for example for $^{+}CMe_3$ $(\delta = -135.4 \text{ p.p.m.})$. Further, cyclopropyl carbinol (14) and cyclobutanol (15) in SbF_5/SO_2FCl solution at $-80°C$ are found to give a common ion for which δ ^{13}C (-137.9 p.p.m.) and $J_{13_{CH}}$ (180 Hz) are identical for all three CH_2 groups indicating a symmetrical ion such as (16)† (Olah, 1970).

(14)   (15)   (16)

† Alternative representations of this ion are outlined by Olah, (1970).

The observed products of displacement reactions of cyclobutyl (and cyclopropyl methyl) derivatives may therefore be rationalised as the result of reaction of the non-classical carbonium ion (16) at alternative sites:

(16)

CH_2Y ⟵ Y^- ⟶ $CH_2{=}CH{-}CH_2{-}CH_2Y$

↓

Y

3.4. Reduction of cycloalkanones.   The sequence of relative rates of borohydride reduction of cycloalkanones:

$$(CH_2)_{n-1}\ C{=}O \xrightarrow[\text{(ii) } H_2O]{\text{(i) } NaBH_4} (CH_2)_{n-1}\ C\begin{smallmatrix}OH\\H\end{smallmatrix}$$

(cf. Table 3.2) is effectively the inverse of that for acetolysis of the cyclo-alkyl toluene-*p*-sulphonates in Table 3.1 (Brown and Ham, 1957).   Over

TABLE 3.2   *Rates ($k_2 \times 10^4$ l/mol s. at 0°C) for reduction of cycloalkanones with sodium borohydride in isopropanol*

n	$k_2 \times 10^4$	n	$k_2 \times 10^4$
4	266	10	0.013
5	7.0	11	0.023
6	161	12	0.182
7	1.02	13	0.194
8	0.08	15	0.42
9	0.03	(n-C_6H_{13})$_2$CO	0.454

the range $n = 5$ to $n = 11$, log k_2 for reduction is in fact linearly and inversely related with log k_1 for the acetolysis of the toluene-*p*-sulphon-ates.   This is understandable since these reactions involve complementary changes, sp$^3 \rightarrow$sp^2, or sp$^2 \rightarrow$sp^3 in co-ordination at the reaction site, and apparently complementary changes in molecular compression in reaching the respective transition states.

A linear relation found between log k_2 for sodium borohydride reduction and log K for the cyanohydrin equilibrium:

$$(CH_2)_{n-1}\ C{=}O \quad \underset{}{\overset{HCN}{\rightleftarrows}} \quad (CH_2)_{n-1}\ C{<}^{CN}_{OH}$$

$$K = [\text{Ketone}]\,[\text{HCN}]/[\text{Cyanohydrin}]$$

for cycloalkanones from $n = 5$ to $n = 17$, illustrates the same point (Sicher, 1962). Significantly, cyclopropyl methyl ketone is relatively slowly reduced by sodium borohydride, and cyclopropyl methyl toluene-*p*-sulphonate is highly reactive. However, the conjugative effect of the cyclopropyl residue (cf. chap. 1) may be important in reducing the reactivity of the keto-group.

3.5. Reactivity of cycloalkenes. The relative reactivities of cycloalkenes are of particular interest in drawing attention to the relative change in steric strain in reaching the reaction transition state, which depends on the particular reaction being considered. This aspect is illustrated in the following examples which compare the reactivities of cyclopentene and norbornene with that of cyclohexene for various reagents (Rocek, 1969).

Reaction	*Transition State*	*Rate relative to cyclohexene*	
+ PhN₃		64	6500
+ B₂H₆		110	—
+ RCO₃H		1.5	1.2
+ Ag⊕		2*	17*

* Stability constants of the Ag⁺–olefin complex.

Strain release increases with the ring size of the cyclic transition state intermediate, i.e. $3 < 4 < 5$-membered. Hence the reactivity of the strained norbornene is relatively greater for the reaction with phenyl azide where there is a five-membered transition state intermediate than with a peracid for example where the intermediate is three-membered.

3.6. Enolisation of cycloalkanones. A similar phenomenon which emphasises the importance of the reaction transition state is encountered in the relative rates of enolisation of cycloalkanones under acidic and basic conditions.

There is substantial evidence that in acid catalysed enolisation:

the transition state resembles the enol. Hence for acid catalysed bromination, for which enolisation is rate limiting, the relative rate order is: cyclobutanone 1, cyclopentanone 150, cyclohexanone 793, cycloheptanone 101, which reflects the ease of introducing an olefinic bond into these rings.

In base catalysed enolisation the transition state is more ketone-like:

and for base catalysed hydrogen deuterium exchange a different relative rate order: cyclobutanone 24, cyclopentanone 7, cyclohexanone 1, cycloheptanone 2, is observed (Schechter *et al.*, 1962).

3.7. Transannular reactions. The conformation taken up by the medium-sized rings ($n = 8$ to 11) results in a number of hydrogen substituents being turned inside the ring (cf. chap. 2). This structural feature leads to the phenomenon of transannular hydride shift in displacement reactions of the general form:

(e.g. $X = OTs$, N_2^+)

An instance of this type of reaction which has been very fully examined is that of cycloalkene oxides with formic acid which leads to the formate

of the transannular diol (18), as well as to the expected formate of the vicinal glycol (17) (together with cycloalkenols, from proton elimination, and in some cases also bicyclic products). Table 3.3 (Cope, 1966) summarises the results of formolysis of cycloalkene epoxides. Slight transannular hydrogen transfer is observed in six- and seven-membered rings. In the range $n = 8$ to 10 it is a major reaction. It is important to

TABLE 3.3 *Glycol products derived from formolysis of cycloalkene oxides*

n	oxide	1,2-Diol, %	Transannular diol,	%
6	*cis*	main product	*trans*-1,4-diol,	0.03
7	*cis*	main product	*cis*-1,4-diol,	2.4
8	*cis*	20%	*cis*-1,4-diol	30
8	*trans*	—	{ *trans*-1,4-diol *trans*-1,3-diol	35
9	*cis* and *trans*	—	1,5-diols	mainly
10	*cis*	—	*cis*-1,6-diol	mainly
10	*trans*	—	*trans*-1,6-diol	mainly

note that the transannular reaction is stereospecific just as is the ring opening of an epoxide to the 1,2-diol:

The transferred hydrogen may be identified. The dideuterocyclo-octene oxide (20) gave the deuterated diols (21), (22) and (23), indicating

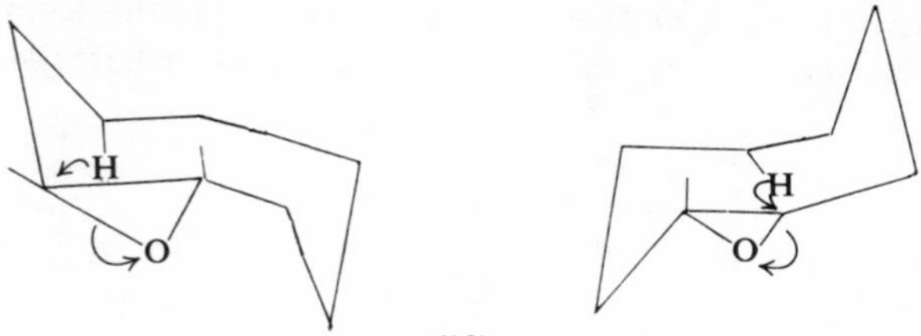

(19)

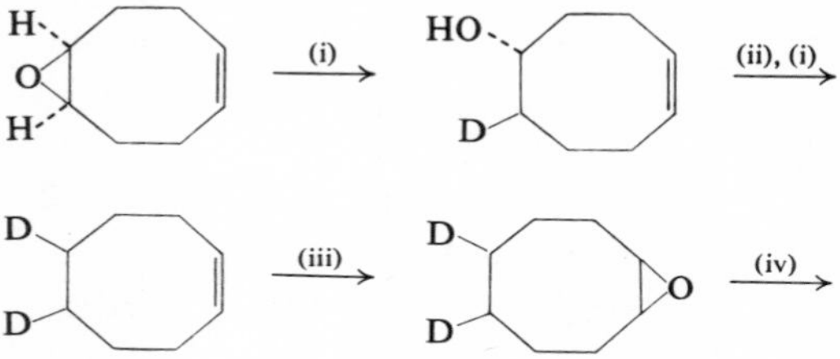

(20)

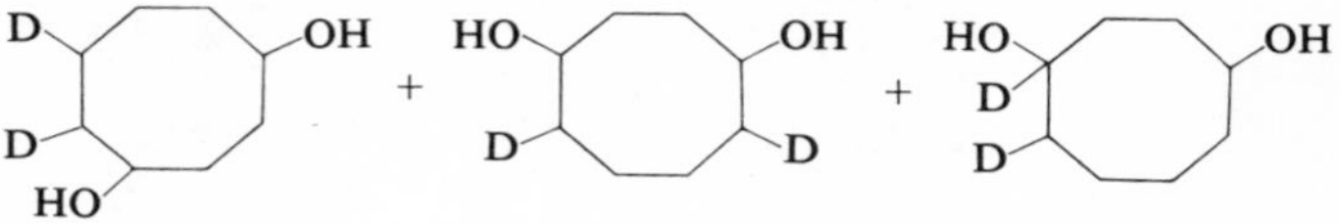

(21)   (22)   (23)

(i) $LiAlD_4$, (ii) p-$BRC_6H_4SO_2Cl$, (iii) monoperphthalic acid, (iv) 90% formic acid.

transannular hydrogen shifts from the third or fifth carbon atom from the reaction site, as the conformation (19) requires.

The 9-, 10-, and 11-membered cycloalkene oxides gave little 1,2-diol, but with cyclododecene oxide reaction again returns to a more normal pattern in giving only 1,2-diol. Cyclo-octene oxide is intermediate in showing both types of reaction, and in this case the relative proportions of normal and transannular reactions may be influenced also by the experimental conditions. A strong, and correspondingly weakly nucleophilic acid, gives more transannular products, viz: cyclo-octene oxide + RCO_2H gives 100, 94 and 87 per cent transannular products when $R = CF_3$, CCl_3 or H respectively. Use of a buffered solution gives much more of the normal vicinal diol and in buffered acetic acid the proportion of transannular product falls to 24 per cent.

Similar evidence of transannular reaction has been derived in the acetolysis of cyclo-octyl toluene-*p*-sulphonate, and in the deamination of cyclodecylamine. The cyclodecyl carbonium intermediate (24) reacts as follows:

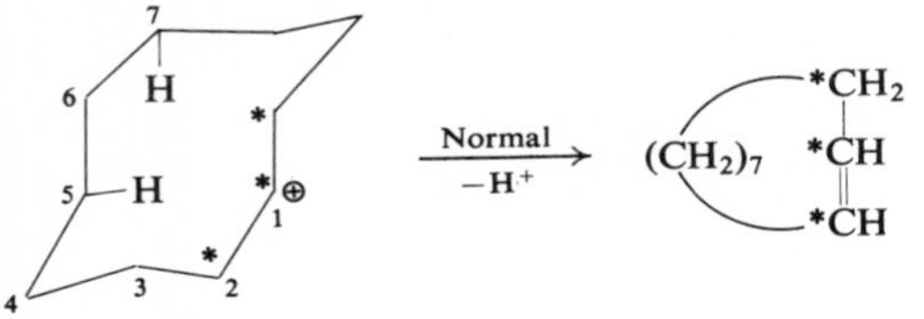

(24)

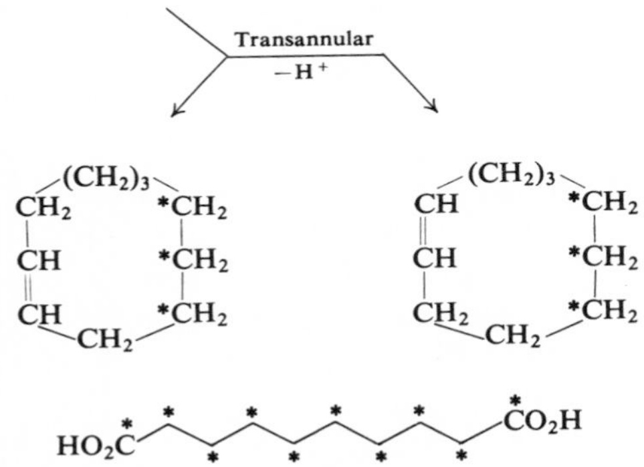

(25)

(* = ^{14}C)

The redistribution of ^{14}C over carbon atoms 4, 5, 6, 7 and 8, as a result of transannular hydrogen transfer, was established by oxidation to sebacic acid (25) and stepwise degradation.

3.8. Transannular reaction and ring conformation.   Transannular hydrogen transfer must, however, clearly depend on the precise ring conformation. This is illustrated by the behaviour of 5,5,8,8-tetramethyl cyclodecene-*cis*-oxide (26) which is hydrated to give the normal diol (27) (Sicher *et al.*, 1965). The corresponding tetramethyl-*cis*-cyclodecene (28) also reacts normally with bromine to give the dibromide (29), whereas the unsubstituted *cis*- and *trans*-cyclodecenes give respectively

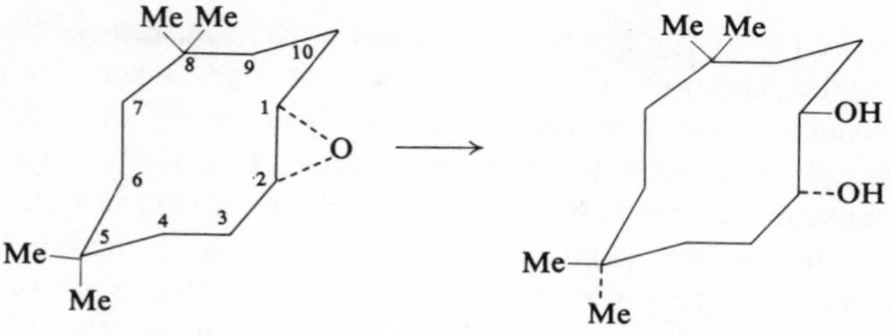

(26)

(27)

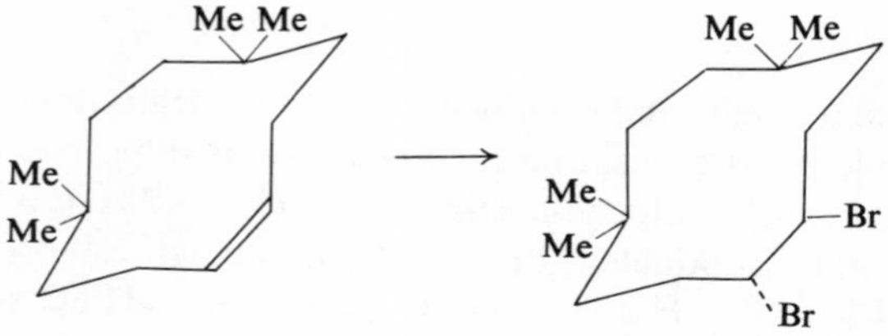

(28)

(29)

cis- and *trans-*1,6-dibromo cyclodecane (Sicher *et al.*, 1965). Trans-annular reaction is not observed with (26) or (28) since the methyl substituents cause twisting of the ring and movement of the transannular hydrogen away from the reaction site. For the same reasons 5,5-dimethyl cyclononyl toluene-*p*-sulphonate is found to be solvolysed without transannular reaction.

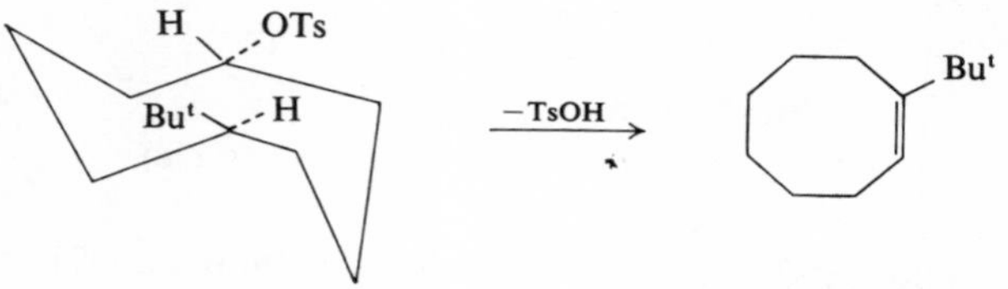

$\xrightarrow{-\text{TsOH}}$

(30)

(31)

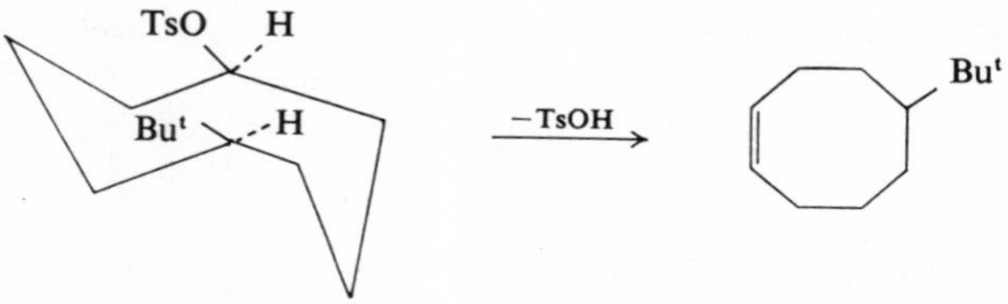

$\xrightarrow{-\text{TsOH}}$

(32)

(88%)

In (26) the methyl substitution is vicinal to the site of possible trans-annular hydrogen transfer. When the substitution is actually on the carbon atom transannular to the reaction site the effect is different. In *cis*-t-butyl cyclo-octyl toluene-*p*-sulphonate (30) the t-butyl group assists transannular reaction so that by solvolysis (31) is obtained in high yield. In the *trans* isomer (32), on the other hand, the alkyl group prevents rearward transfer of a transannular hydrogen so that (32) gives mainly the normal product of a 1,2-elimination reaction (Allinger and Greenberg, 1962).

3.9. Transannular cyclisation reactions. When the transannular group-ing is an olefinic bond cyclisation may occur to give bicyclic products, e.g. the cyclodecenyl toluene-*p*-sulphonate (33)→(34) (and smaller amounts of other products), or the cyclo-octenyl ester (35)→(36; R = H, R' = HO; or R = HO, R' = H) (Cope, Grisar and Peterson, 1960).

(33)   (34)

(35)   (36)

Solvolysis of cyclohept-3-enyl bromobenzene-*p*-sulphonate, (37), which is an example of homallylic bond participation in the displacement reaction leads to the bicyclo[4,1,0]heptanol (38) (Cope, Park and Scheiner, 1962). The importance of these reactions yielding bicyclic

(37)   (38)

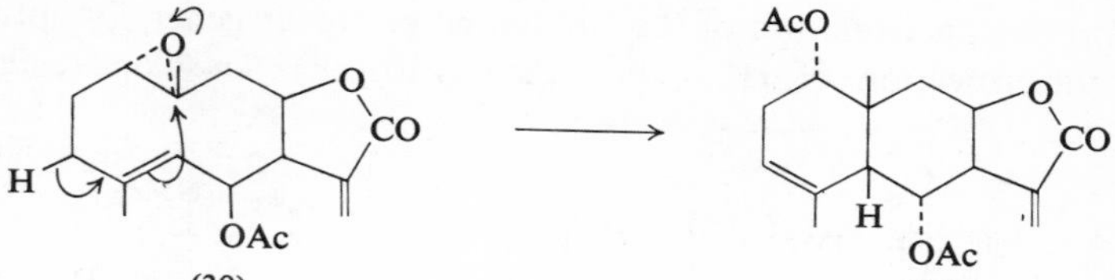

(39)   (40)

structures is also well illustrated in the chemistry of terpenes containing medium-sized rings. For example, pyrethrosin (39), which is an epoxy-cyclodecene, is cyclised to (40) (Barton *et al.*, 1960).

3.10. Transannular interaction and ultraviolet absorption. Transannular interaction is seen also in the ultraviolet absorption of (42) in comparison with (41). In (42) the proximate olefinic group moves the carbonyl absorption to longer wavelengths and there is some increase in absorp-

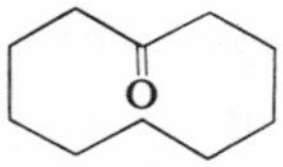

(41)   λ 288 nm, ε, 16

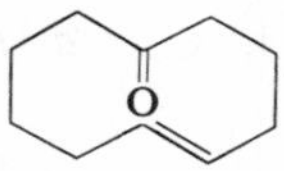

(42)   λ 302 nm, ε, 73

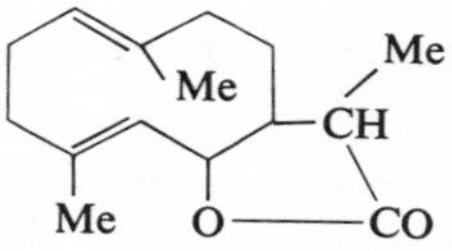

(43)

tion intensity. Similarly in the terpene (43) the proximate olefinic groups lead to very intense end absorption in the ultraviolet.

3.11. Reaction of vicinal groups, the Hofmann and Cope eliminations. In the Hofmann elimination:

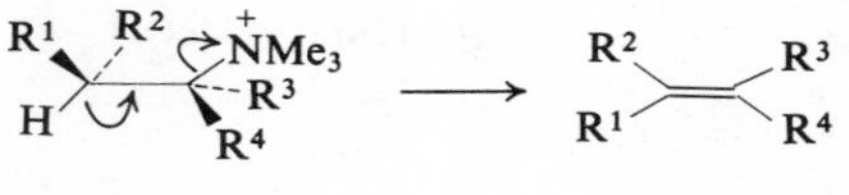

(44) 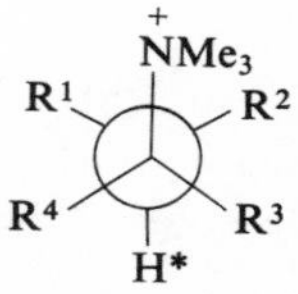   (46)

an *anti*-coplanar relation of the eliminated groups is generally optimal, as in the projection (46, H* = eliminated hydrogen). The Cope reaction:

(45)   (47)

on the other hand, exhibits the complementary steric requirement of near coplanarity of *syn* related groups, cf. projection (47, H* = eliminated hydrogen), cf. Banthorpe (1963).

For the smaller rings (*n* = 6 or 7) both reactions lead to the *cis*-cycloalkene (cf. Table 3.4), although in the Hofmann reaction the yield

TABLE 3.4   *Cycloalkenes from cycloalkylamines*

	Method			Method	
n	Hofmann (%)	Cope (%)	*n*	Hofmann (%)	Cope (%)
6	*cis* (62)	*cis* (93)	9	*trans* (89)	*trans* (90)
7	*cis* (87)	*cis* (92)	10	*trans* (90)	*trans* (94)
8	*cis* and *trans*	*cis* (—)			

of cycloalkene may be reduced owing to the competing reaction:

This arises since to establish the relationship shown in (46) the $-\overset{+}{\text{N}}\text{Me}_3$

group must necessarily be axially oriented, which requires preliminary ring inversion, i.e.:

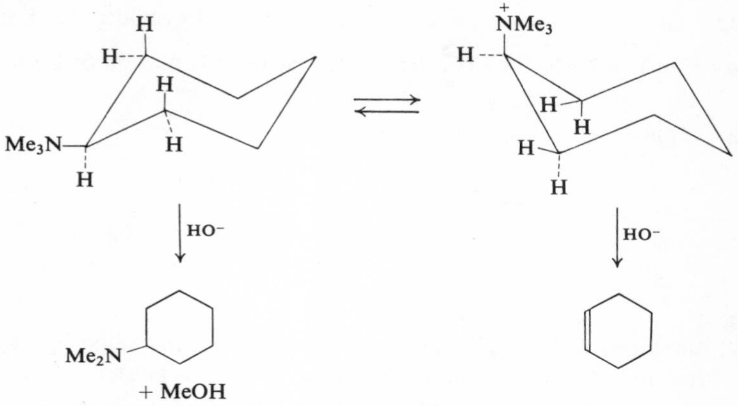

In the larger rings, however, the *trans*-cycloalkene is the normal product of Hofmann or Cope elimination (Sicher *et al.*, 1966; 1967). The reason for this change in stereochemistry, i.e. from *cis* ($n = 6$ or 7) to *trans* ($n = 9$ or 10), with the cyclo-octyl case giving a mixed product, may be seen by comparing (46) or (47) with (48) which represents the situation in a larger ring. The steric situation shown in (48) arises because the larger

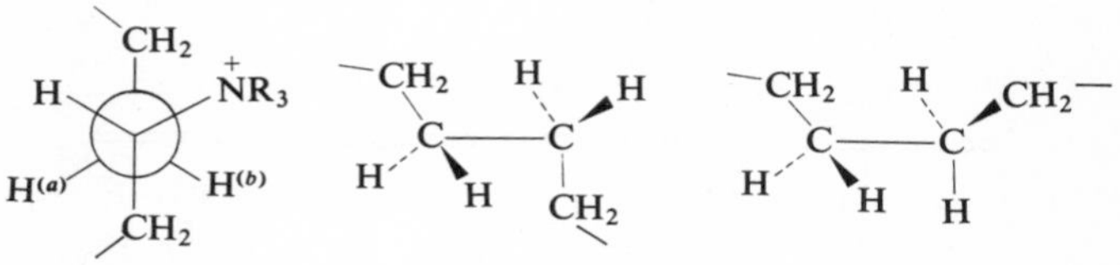

rings may accommodate a $-CH_2-CH_2-CH_2-CH_2-$ unit close to the *anti*-conformation, i.e. as in (49), whilst in the smaller rings these butane units are necessarily near to the gauche arrangement (50). In (48) Hofmann elimination of an NMe_3 group with $H^{(a)}$, or Cope reaction of an Me_2N-O group with $H^{(b)}$ both give the *trans*-cycloalkene. The cyclo-octyl derivatives represent an intermediate case between (46) and (48).

3.12. *Syn-* and *anti*-elimination in medium-sized rings. Close study of elimination reactions in medium-sized rings has also provided important information regarding reaction mechanisms. Examination of the base-catalysed elimination reactions of the tetramethyl cyclodecane deriva-tives (51, Y = $\overset{+}{N}Me_3$ or OTs) has indicated that formation of both *cis-*

(51)   (52)   (53)

(52) and *trans*-olefins (53) involves elimination of the same hydrogen substituent, $H^{(b)}$ (Sicher *et al.*, 1966; 1967). This implies a *syn*-elimination as well as the better known *anti*-elimination process, which although generally optimal is evidently not always obligatory.

Direct evidence of *anti*-oriented units in nine-membered and larger rings is provided by the molecular dipole moment of a series of lactones, cf. Table 3.5 (Huisgen and Ott, 1959). The smaller ring lactones show a

TABLE 3.5   *Molecular dipole moments of lactones*

n	μ,	Debye units	n	μ,	Debye units
4	3.80	*syn*	8	3.70	*syn* and *anti*
5	4.09	*syn*	9	2.25	*anti* and *syn*
6	4.22	*syn*	10	2.01	*anti*
7	4.45	*syn*	11	1.86	*anti*

dipole moment close to the value of *c.* 4.3 D characteristic of the *syn*-orientation (54). The nine-membered lactone evidently represents a point of change over towards a moment corresponding to the *anti*-orientation (55). The difference between (54) and (55) arises from the different orientation of the bonds and hence also of the non-bonding electrons around the oxygen relative to the $\diagup$C$=$O bond dipole.

(54), μ 4.3 D   (55), μ 1.77 D

Although the steric requirements of the alkane unit, $-CH_2-CH_2-$, and the lactonic group, $-CO-O-$, cannot be completely equated there is nevertheless a significant parallel between the data of Tables 3.4 and 3.5.

3.13. Reduction of medium-sized ring cycloalkynes.   The possibility of accommodating a more nearly linear four-carbon unit in the larger rings also lies behind the occurrence of the cycloalkynes (cf. chap. 1) and their behaviour towards sodium/liquid ammonia/sodium amide.   The tetramethyl cyclodecyne (56) is an illustrative case.   Slow reduction to the

trans-olefin (57) is accompanied by isomerisation catalysed by the sodium amide present to the allene (58), and thence rapid reduction to the *cis*-olefin (59) which forms the major product (Svoboda *et al.*, 1964).

References

Allinger, N. L. and Greenberg, S. (1962). *J. Amer. Chem. Soc.*, **84**, 2394.

Banthorpe, D. V. (1963). *Elimination Reactions*, Elsevier.

Barton, D. H. R., Böckman, O. C. and de Mayo, P. (1960). *J. Chem. Soc.*, 2263

Brown, H. C. and Ham, G. (1956). *J. Amer. Chem. Soc.*, **78**, 2735; (1957). *Tetrahedron*, **1**, 221; 231.

Cope, A. C., Grisar, J. M. and Peterson, P. E. (1960). *J. Amer. Chem. Soc.*, **82**, 4299.

Cope, A. C., Park, C. H. and Scheiner, P. (1962). *J. Amer. Chem. Soc.*, **84**, 4862.

Cope, A. C. (1966). *Quarterly Reviews*, **20**, 119.

Eliel, E. L. (1962) *Stereochemistry of Carbon Compounds*, McGraw-Hill, pp. 265–9.

Foote, C. S. (1964). *J. Amer. Chem. Soc.*, **86**, 1853.

Huisgen, R. and Ott, H. (1959). *Tetrahedron*, **6**, 253.

Olah, G. *et al.* (1965). *J. Amer. Chem. Soc.*, **87**, 2998; 5123; (1970). *ibid.*, **92**, 2544.

Roberts, J. D. (1951). *J. Amer. Chem. Soc.*, **73**, 2509; (1959) *ibid.*, **81**, 4390.

Rocek, J. (1969). *J. Amer. Chem. Soc.*, **91**, 991.

Schechter, H. *et al.* (1962). *J. Amer. Chem. Soc.*, **84**, 2905.

Schleyer, P. von R. (1964). *J. Amer. Chem. Soc.*, **86**, 1854.

Sicher, J. (1962) in *Progress in Stereochemistry*, Ed. de la Mare, P. B. D. and Klyne, W., **3**, 222, Butterworths.

Sicher, J., Svoboda, M. and Vaver, V. A. (1965). *Chem. Comm.*, 12.

Sicher, J. *et al.* (1966). *Tetrahedron Letters*, 1619; 1627.

Sicher, J. *et al.* (1967). *Chem. Comm.*, 394.

Svoboda, M., Sicher, J. and Savada, J. (1964). *Tetrahedron Letters*, 15.

4 Cyclisation, cycloaddition and electrocyclic reactions

4.1. The energy and entropy of cyclisation. Cyclisation of a chain of atoms, e.g. $XCH_2CH_2 \ldots CH_2Y$, depends not only on the reactivity of the terminal CH_2 groups, conferred by the substituents X and Y, but there is also the kinetic requirement that these terminal groups should be brought within bonding distance, and with suitable activation energy.

An aliphatic carbon chain has considerable rotational freedom which increases with the chain length. Cyclisation reduces this rotational freedom and, in consequence, the probability of ring closure falls with increasing ring size, i.e. with increasing ring size cyclisation has an increasingly negative entropy of activation.

The activation energy of cyclisation is associated not only with the process of bond breaking and bond making at the reaction sites, but also with the bond–bond opposition forces and changes in carbon–carbon bond angle introduced in the cyclic structure. In forming the smallest rings there is the energy of compression of the normal carbon–carbon bond angle, and in the medium-sized (8–10) rings there is bond-angle widening. Also, formation of the small near-planar rings introduces bond eclipsing forces, and torsional strain raises the energy of the medium ring carbocycles.

The following data (Capon, 1964) for acid catalysed cyclisation of γ-hydroxybutyric acid (1) and δ-hydroxyvaleric acid (3):

$$HOCH_2.CH_2.CH_2.CO_2H \rightarrow$$
(1)
ΔS^+ $-48 J/$deg mole
ΔH^+ 80 kJ/mole

(2)

$$HOCH_2.CH_2.CH_2.CH_2.CO_2H \rightarrow$$
(3)
ΔS^+ -103 J/deg mole
ΔH^+ 58 kJ/mole

(4)

illustrate the energy and entropy requirements for formation of different

63

sized rings. Cyclisation to form the smaller γ-lactone ring (2) has the smaller entropy requirement, but the larger energy of activation.

4.2. Cyclisation: yield v. ring size. Preparation of the relatively strain-free five- or six-membered rings generally presents no difficulty, but the strain energy of three- or four-membered rings necessarily require cyclisation methods based on irreversible reactions. For the medium-sized and large rings the problem is to bring the terminal atoms together, i.e. to effect intramolecular cyclisation rather than intermolecular condensation of a number of chain units into polymeric products.

The problem of ring synthesis is illustrated by the data (Sicher, 1962), in Table 4.1 for the yield of cyclic products obtained (*a*) by Dieckmann cyclisation of a diester:

and (*b*) by Zeigler's method of carrying out the Thorpe cyclisation of a dinitrile using a very dilute solution of the dinitrile so as to minimise intermolecular condensation:

The failure of the Dieckmann and Thorpe–Zeigler methods for the medium-sized rings is striking. However, a general method of cyclisation applicable over a wide range of ring sizes surmounts this difficulty. This is the acyloin condensation of a diester to the acyloin (5) at the surface of clean, finely divided metallic sodium in complete absence of oxygen:

(5)

TABLE 4.1 *Yields of cyclic products obtained by (a) Dieckmann, (b) Thorpe–Zeigler, and (c) acyloin condensation (Yield % by cyclisation method)*

n	(a)	(b)	(c)
6	—	95	58
7	47	95	52
8	14	78	36
9	6	0	38
10	0	0	65
11	0	2	66
12	0	—	80
13	22	14	80
14	30	57	85

The yields obtained by this method are listed under the heading of method (c) in Table 4.1. The sodium surface is able to bring together the terminal ester group without too great intermolecular condensation. Possibly cyclisation is fast relative to adsorption.

4.3. Cycloaddition reactions. In suitable cases cyclic products are most readily available by cycloaddition of appropriate units (cf. Huisgen, 1964). Detailed instances are described in subsequent chapters, but the principle of cycloaddition may be illustrated:

(i) Many cyclopropane syntheses are based on cycloaddition of 2 + 1 carbon units, e.g. diazomethane and ketene give cyclopropanone, diazomethane being effectively a source of a methylene unit, CH_2.

$$CH_2{=}C{=}O + CH_2N_2 \longrightarrow \ \triangleright{=}O + N_2$$

(ii) 2 + 2 Cycloaddition to produce a cyclobutane is possible by thermal addition of polyfluoroethylenes (Roberts, 1962):

$$F_2C{=}CFCl \xrightarrow{\ \Delta\ } \quad \begin{array}{c} F_2 \ \ (F)Cl \\ \square \\ F_2 \ \ (F)Cl \end{array}$$

or, for a less reactive olefin, by photochemical activation, e.g.

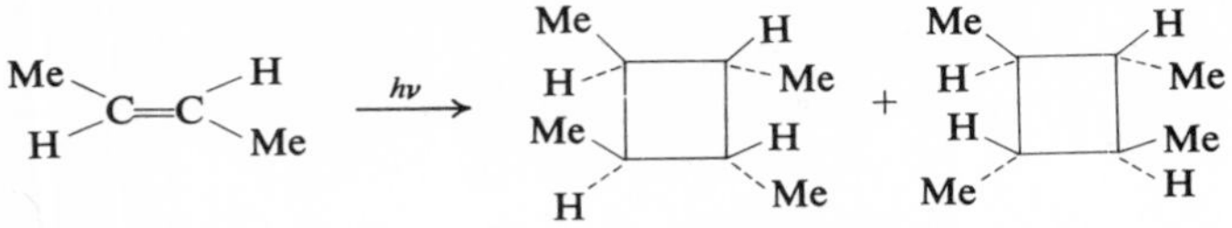

(iii) The Diels–Alder reaction is a particularly valuable instance of cycloaddition of 4 + 2 units e.g.:

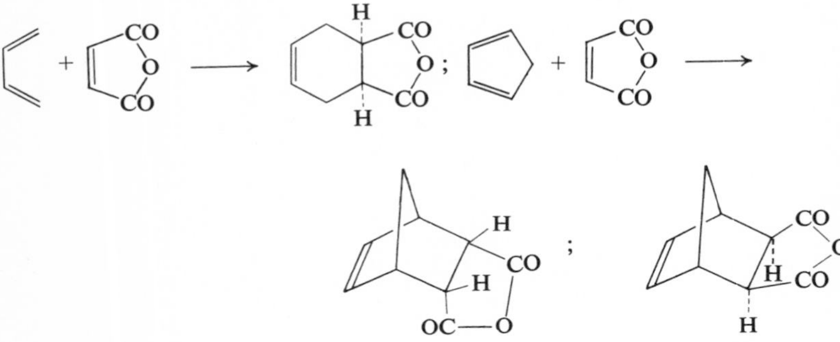

(6)   (7)

and is characterised by preferential formation of the *endo-* (6) rather than the *exo-*oriented product (7).

(iv) Direct synthesis of eight-membered rings is possible *via* suitable nickel complexes of butadiene which undergoes a 4 + 4 cycloaddition (Wilke, 1963, 1970):

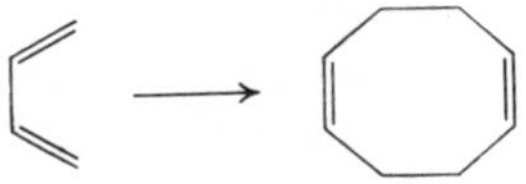

(8)

to give cyclo-octa-1,5-diene, (8), or by 2 + 2 + 2 + 2 tetramerisation of acetylene also *via* a nickel complex to give cyclo-octatetraene, (9), (cf. chap. 7) (Schrauzer, 1964).

$$4 \ HC{\equiv}CH \longrightarrow$$

(9)

4.4. Ring expansion and ring contraction. The availability of cyclo-hexane derivatives makes ring contraction a generally convenient route to cyclopentane derivatives, e.g.

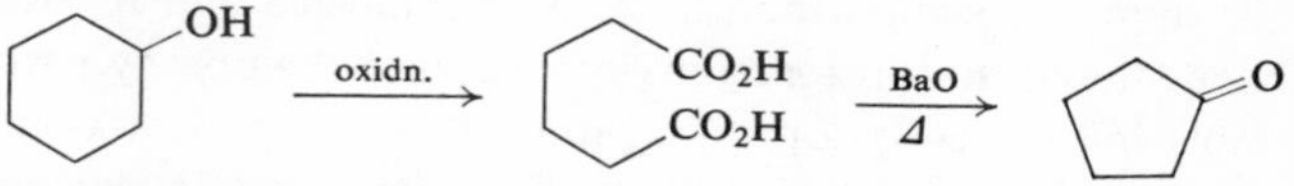

in which cyclohexanol is a source of adipic acid which is ketonised by heating with baryta. The Favorskii reaction (Kende, 1960) of an α-haloketone may also be employed for ring contraction: 2-bromocyclo-

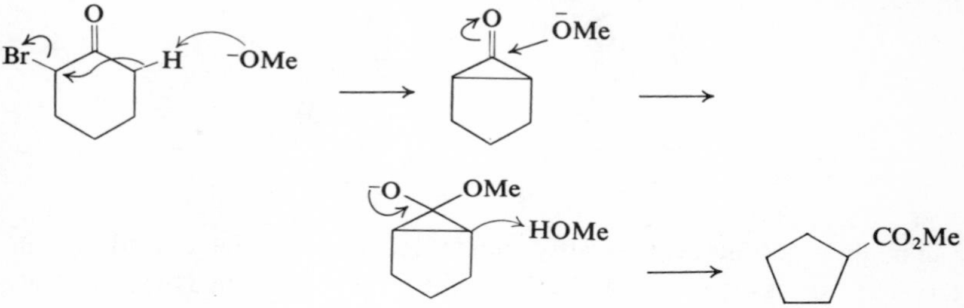

(10)

hexanone giving, e.g. methyl cyclopentane-carboxylate (10).

Similarly cycloheptane derivatives may be obtained by ring expansion (cf. Gutsche and Redmore, 1968) of an appropriate cyclohexane, e.g. cyclohexanone and diazomethane give cycloheptanone (11).

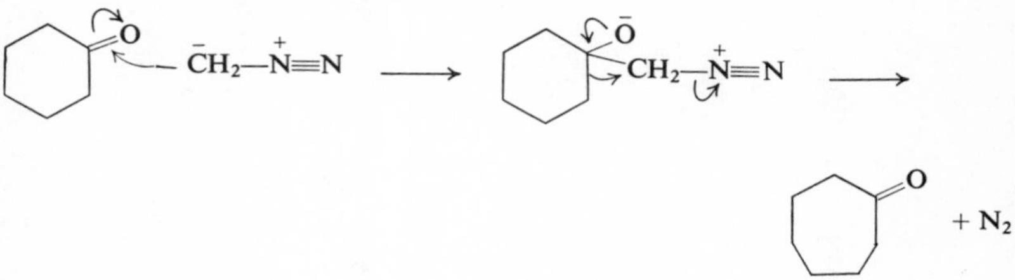

(11)

The Demjanov rearrangement brought about by nitrous acid deamina-tion is also valuable (Smith and Baer, 1960).

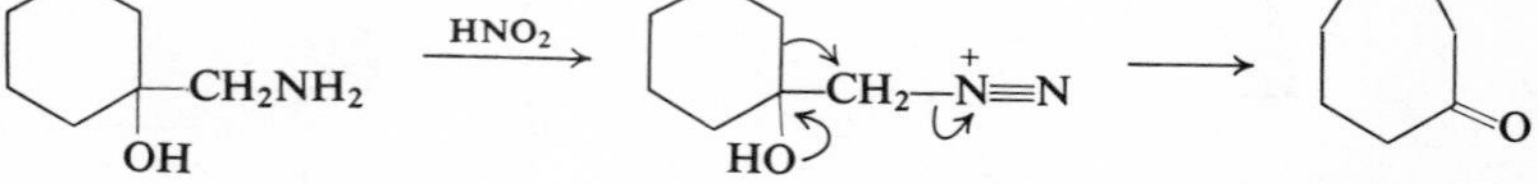

(11)

4.5. Molecular orbital symmetry correlations.   A number of cyclisation processes, e.g. the photo-dimerisation of an olefin to cyclobutane, or Diels–Alder addition, involve reaction between the termini of a π-electron system to form a σ-bond. R. B. Woodward and R. Hoffmann (1969) have shown that in reactions of this kind the energy pathway is considerably influenced by orbital symmetry considerations. Thus (12) represents the orbital overlap in the bonding level of ethylene, and (13) the anti-bonding or excited level, the shaded and unshaded lobes indicating opposite phases of the electronic wave function.

(12)   (13)

The interaction of two ethylene molecules may be considered in terms of an association of the bonding level (12) as in (14) and (15),

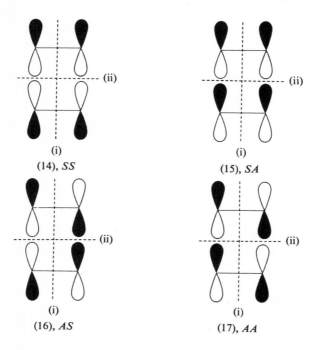

(14), *SS*   (15), *SA*

(16), *AS*   (17), *AA*

and as in (16) and (17) for interaction of the orbitals of the anti-bonding level, (13). The symmetries of these combinations are determined by reflection in planes (i) and (ii). (14), which is symmetric (S) in relation to both planes, is designated an SS combination, and (15), which is symmetric about plane (i) and anti-symmetric (A) about plane (ii), is designated SA. The symmetries of (16) and (17) are as indicated.

Since bonding interaction is dependent on overlap of orbitals of the same phase the situations represented by (14) and (16) are bonding, whilst (15) and (17) are anti-bonding.

The appropriate bonding orbitals for the cyclobutane product of cycloaddition are shown at (18) and (19), with the corresponding anti-bonding levels at (20) and (21). The symmetries relative to planes (i) and (ii) are indicated.

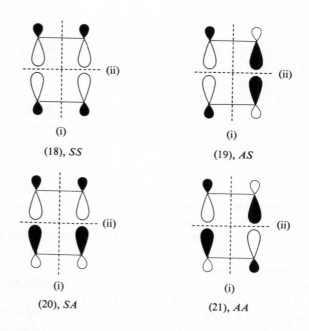

(18), SS (19), AS

(20), SA (21), AA

Inspection of the ethylene combinations (14)–(17) shows correlation of symmetries between (14) and (18), (16) and (19), and of (15) with (20) and (17) with (21). These correlations may be summarised:

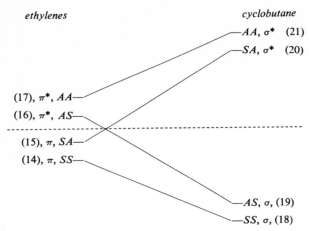

where — represents energy levels relative to the non-bonding level indicated - - - -.

This symmetry correlation diagram indicates clearly that thermal cyclisation of ethylene is an energy-demanding process leading to cyclobutane in an excited state, i.e. (14) and (15) correlate with the σ- (18) and σ*-levels (20) of cyclobutane.

Conversely, it is clear that thermal decomposition of cyclobutane starting from the σ-levels (18) and (19) is directed by symmetry requirements to give ethylene in a π* excited energy level (16). Thus the reaction:

$$\text{cyclobutane} \rightarrow 2\ CH_2{=}CH_2$$

is a process of high activation energy. This conclusion may also be expressed by saying that what might have been considered an easy transformation of cyclobutane into two molecules of ethylene is symmetry disallowed. Thus orbital symmetry considerations are concerned with the pathways of reactions, and enable a distinction to be drawn between symmetry permitted processes of relatively low activation energy, and cases where symmetry relations introduce a considerable energy requirement, i.e. the reaction is symmetry forbidden.

These considerations are, however, relevant only to concerted reaction processes, and do not apply, for example, to the dimerisation of a polyfluoroethylene, which is stepwise through a radical intermediate:

e.g. $F_2C{=}CF_2 \longrightarrow \begin{array}{c} F_2\dot{C}{-}CF_2 \\ | \\ F_2C{-}\dot{C}F_2 \end{array} \longrightarrow \begin{array}{c} F_2 \quad F_2 \\ \square \\ F_2 \quad F_2 \end{array}$

and although symmetry forbidden the decomposition of cyclobutane:

$$\square \xrightarrow{500°} 2\ CH_2:CH_2$$

is in fact observed at a sufficiently high temperature.

The symmetry-imposed barrier to the thermal interconversion:

$$2\ C_2H_4 \rightleftharpoons cyclobutane$$

does not, however, extend to the photochemical cycloaddition. It was noted above that but-2-ene may be dimerised photochemically to tetra-methyl cyclobutanes. The feasibility of the photo-reaction *via* an excited state is clear from the correlation diagram which relates olefin π- (14), and π^*-levels, (16), with the σ-levels, (18) and (19), of cyclobutane.

4.6. Molecular orbital symmetries and Diels–Alder addition. Symmetry considerations apply also to the Diels–Alder $4 + 2$ cycloaddition. For reaction the diene and olefinic components must be brought together as in (22) so as to permit overlap of the π-electron lobes. This association is characterised by a single plane of symmetry (a) bisecting the two reactant molecules. The occupied orbitals of the olefin and diene are shown

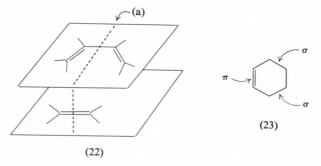

(22)   (23)

in (24), (25), and (26), the symmetry of each relative to the plane (a) being indicated.

The two new σ- and one π-bond generated in forming the cyclohexene product are indicated at (23), and the corresponding orbitals and their symmetry relative to the plane (a) are shown at (27), (28), and (29). The diagram indicates the direct symmetry correlation of the orbitals of the reactants and the product, i.e. Diels–Alder addition is a concerted thermal process which is symmetry allowed.

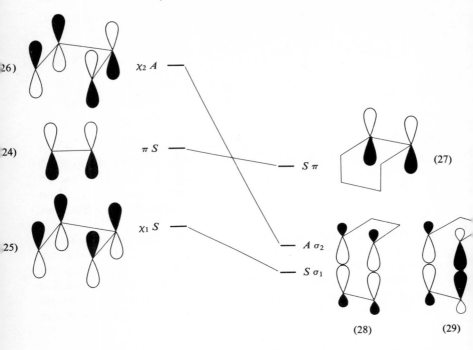

Symmetry considerations may also account for the well known preference for *endo*-addition in the Diels–Alder reaction, as noted in the instance
(6) above. This is ascribed to weak secondary interactions between
occupied and unoccupied orbitals in the diene and dienophile which
Woodward and Hoffmann (1970) illustrate with reference to the interaction of two molecules of butadiene which gives 4-vinyl cyclohexene
(30) as the main reaction product:

Endo-association involving either the highest occupied diene level with
the lowest unoccupied level of the dienophile as in (31), or the lowest
unoccupied diene and highest occupied dienophile levels (32) brings

together on atoms 3 and 3' orbitals of suitable symmetry for overlap interaction.

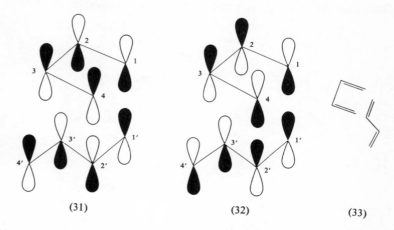

(31)   (32)   (33)

This secondary bonding cannot arise in the *exo*-association shown in (33). However, Diels–Alder addition is a reversible process, and by dissociation into the components, *endo-* → *exo*-isomerisation is possible.

Thus cyclopentadiene with maleic anhydride gives the *endo*-bicyclo-heptene dicarboxylic anhydride (34) which is the kinetic product, but the stable *exo*-isomer (35) is produced at temperatures above 165°.

4.7. Stereospecificity in electrocyclic reactions. Orbital symmetry considerations are important also in understanding the stereospecificity observed in the thermal decomposition of cyclobutenes:

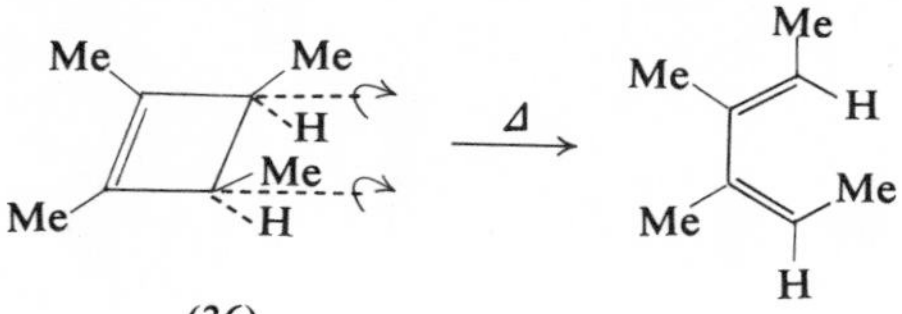

(36)

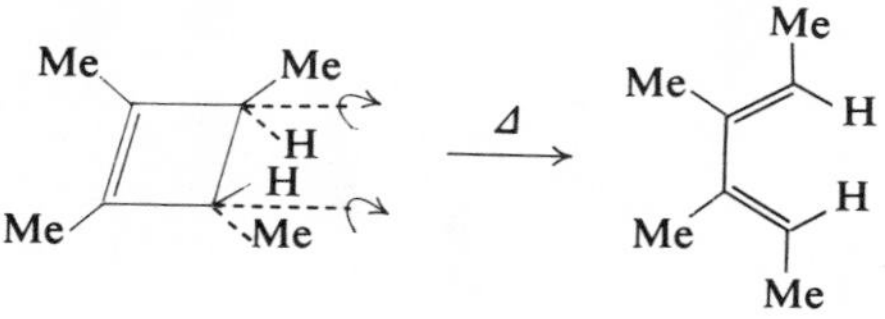

(37)

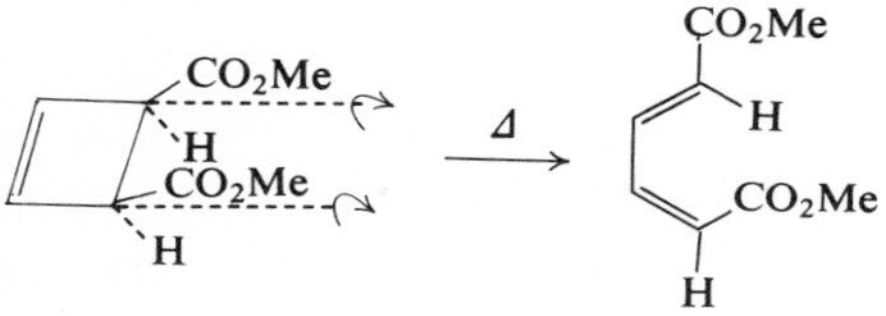

(38)

or of the thermal, and photo-cyclisation of the substituted hexa-1,3 *cis*, 5-triene (39) to different cyclohexadienes (40) and (41).

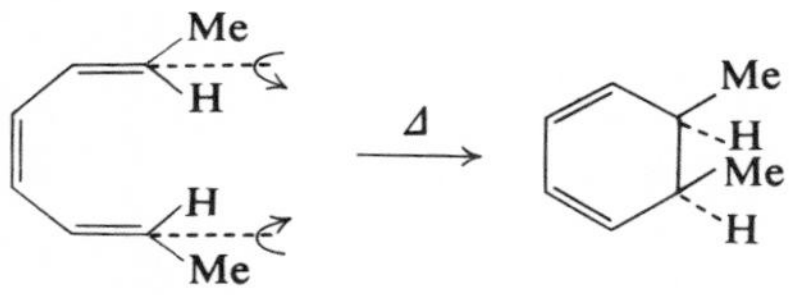

(39)   (40)

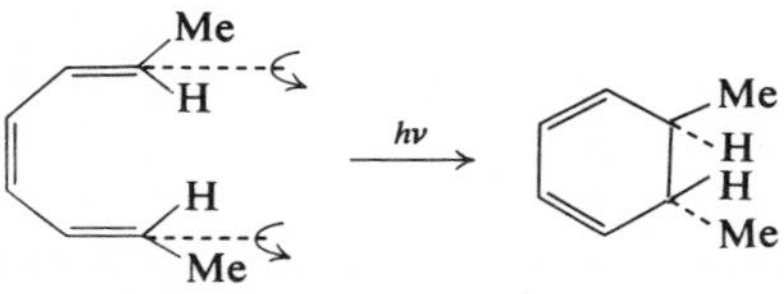

(39)   (41)

These represent what Woodward and Hoffmann call electrocyclic reactions. The stereochemistry of the products from the cyclobutenes (36), (37), and (38) is derived by a process of rotation of the substituent groups in the same sense, as indicated by the arrows. In the cyclisation of the hexatriene (39) the thermal (40) and photochemical (41) products are derived, as indicated by rotatory movements of substituents in opposite senses in the two cases.

Processes of rotation in the same, or in opposite senses have been named conrotation, cf. (42), and disrotation, cf. (43), respectively:

(42)

(43)

The stereospecificity of the cyclobutene→butadiene transformation illustrated by (36), (37), and (38), or of the hexatriene cyclisations, (39)→(40) or (41), may be deduced most directly by considering the symmetry of the highest occupied orbital of the diene or triene component. Inspection of the highest occupied orbital of butadiene (44) shows that to bring together terminal orbital lobes of the same phase requires conrotation which must therefore also characterise the converse transformation of a cyclobutene to a butadiene. Similarly, the highest occupied orbital for the hexatriene (45) indicates the necessity for disrotation to bring together terminal lobes of the same phase in forming a cyclohexadiene. The requirement for conrotation in the photocyclisation of a hexatriene i.e. (39)→(41), is rationalised by reference to the orbital symmetry properties of the lowest excited level (46) which will be concerned in the photoreaction.

These conclusions regarding the basis of the stereospecificity of these electrocyclic reactions may be fully confirmed by a complete symmetry correlation between all the orbitals of the reactants and the products.

A further important example of orbital symmetry control of a reaction pathway is seen in the solvolysis of a cyclopropyl derivative (47, e.g. X = OTs) *via* the allyl cation (48) which is the reaction intermediate in this case. Solvolysis is dependent on mobilisation of the electrons

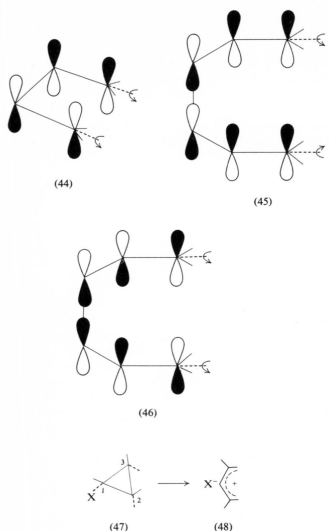

(44)

(45)

(46)

(47)   (48)

of the 2,3-bond for rearward displacement of X as shown in (49) by the
arrow (i). This process in turn necessitates a process of disrotation about
the 1,2- and 1,3-bonds as shown by the arrows (ii). The requirement
for disrotation in the sense necessary for rearward displacement of X

lies behind the considerable difference in rate of solvolysis of *endo*-
(50) and *exo*- (52), norcarane toluene-*p*-sulphonates, viz.: *endo/exo* =
10^4. Inspection of the intermediate allylic cations (51) and (53) clearly
relates slow solvolysis of (52) with the highly unfavourable transoid

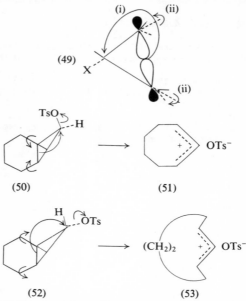

arrangement (53) derived from (52) by a process of disrotation. Further
examples are considered in connection with the chemistry of cyclopropanes in chap. 5.

The requirement for con- or disrotation may be related to the number
of π-electrons concerned. The allyl cation, $\overset{\displaystyle |}{\underset{\displaystyle }{>}}C=\overset{+}{C}-C<$ represents
two π-electrons, the butadiene ⇌ cyclobutene case involves four, and
the hexatriene ⇌ cyclohexadiene case six π-electrons. This may be
generalised for thermal and photo reactions:

Number of π-electrons	Thermal reactions	Photochemical reactions
$4n$	Conrotatory	Disrotatory
$4n + 2$	Disrotatory	Conrotatory

where $n = 0, 1, 2 \ldots$

References

Capon, B. (1964). *Quarterly Reviews*, **18**, 45.
Gutsche, C. D. and Redmore, D. (1968). *Carbocyclic Ring Expansions*, Academic Press.
Huisgen, R. (1964) in *The Chemistry of Alkenes*, Ed. Patai, S., Interscience, pp. 755, 778ff.
Kende, A. S. (1960). *Organic Reactions*, **11**, 261.
Roberts, J. D. (1962). *Organic Reactions*, **12**, 1.
Schrauzer, G. N. (1964). *Angewandte Chemie* (Internat. Edn.), **3**, 185.
Sicher, J. (1962) in *Progress in Stereochemistry*, Ed. de la Mare, P. B. D. and Klyne, W., Butterworths, **3**, 222.
Smith, P. A. S. and Baer, D. R. (1960). *Organic Reactions*, **11**, 157.
Wilke, J. (1963). *Angewandte Chemie* (Internat. Edn.), **2**, 105; (1970). *Advances in Organometallic Chemistry*, **8**, 29.
Woodward, R. B. and Hoffmann, R. (1969). *Angewandte Chemie* (Internat. Edn.), **8**, 781; cf. Woodward, R. B. and Hoffmann, R. (1970). *Conservation of Orbital Symmetry*, Verlag Chemie/Academic Press.

5 The chemistry of small rings

5.1. Cyclopropanes. The chemistry of cyclopropanes reflects the relative bond orders, cf. (1) and chap. 1, of the C–C and external bonds, which retard displacement processes of the kind of (2, X = hal., OTs), and conversely assist reactions of the kind (3) or (4), and also assist ring fission by an electrophile as in (5).

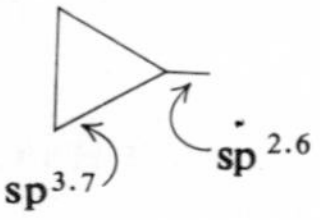

(1) 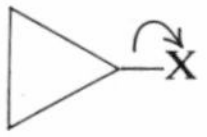 (2)  (3) 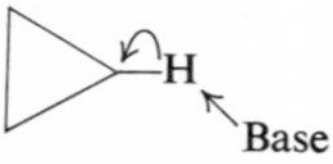 (4) 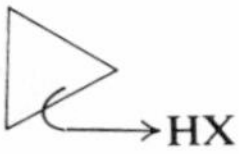 (5)

$$CH_3.CH_2.CHO$$

The effect of bond order on displacement reactions of the type (2) is well illustrated by the following data for the rates of acetolysis of toluene-*p*-sulphonates relative to cyclohexyl toluene-*p*-sulphonate: isopropyl toluene-*p*-sulphonate, 1.4; cyclopropyl toluene-*p*-sulphonate, 2.1×10^{-5}. Clearly fission of the cyclopropyl toluene-*p*-sulphonate bond is seriously retarded. The displacement in fact proceeds with rearrangement, *via* the allyl carbonium ion (6) to give the allyl ester (7). i.e. the

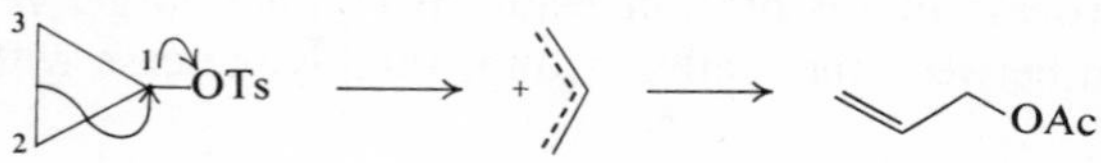

(6) (7)

electrons of the 2,3-bond are engaged, cf. chap. 4. For this reason displacement may be assisted more by alkyl substitution at positions 2 and

3 than at position 1. Thus (8), (9), (10), and (11) show relative rates of acetolysis of 1, 180, 4×10^4 and 500 respectively (Schleyer *et al.*, 1966).

(8)

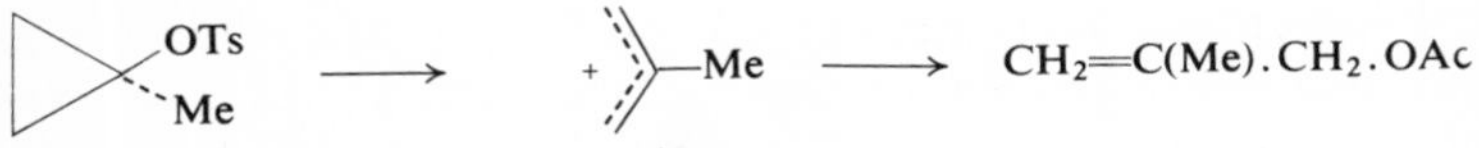

(9)

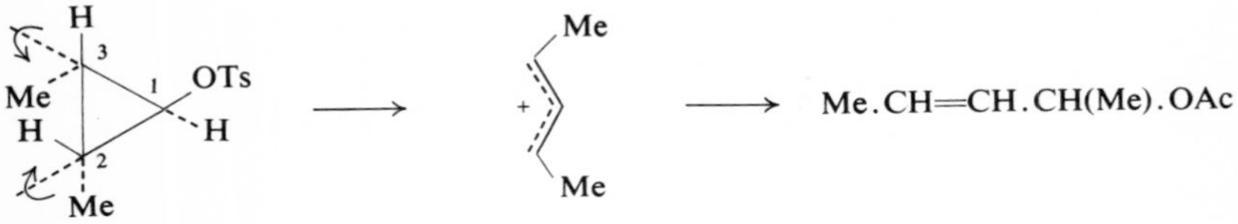

(10)

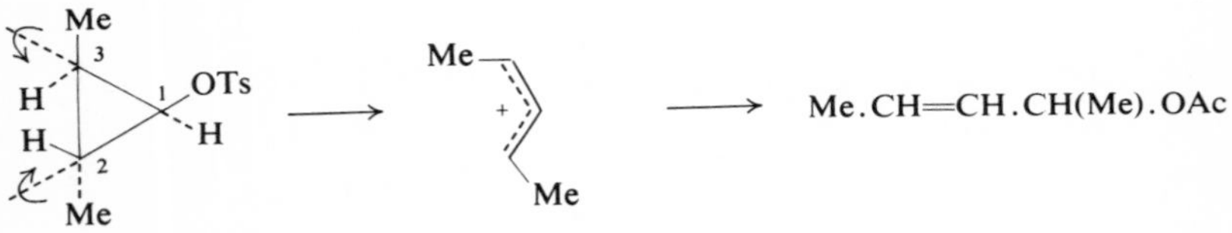

(11)

The difference in reaction rate between (10) and (11) also emphasises the importance of the stereochemistry of 2,3-substitution, owing to obligatory disrotation about the 1,2 and 1,3 bond axes as indicated by the arrows in the sense required for rearward displacement of the leaving OTs group. This depends on the operation of the Woodward and Hoffmann rules relating to orbital symmetries discussed in chap. 4. For this reason (10) reacts the more rapidly since in the *cis*-2,3-dimethyl isomer a process of disrotation leads to a much larger reduction in compression between the methyl groups than is the case with the *trans* isomer (11).

5.2. Cyclopropanols: ring opening. Cyclopropanols readily undergo ring opening. The process is catalysed by acid as in (12), where electron release by the hydroxyl oxygen assists the characteristic cyclopropane ring fission by acid. The importance of this electron release is emphasised, however, by the fact that ring fission is also base catalysed, cf. (13), and is then much faster than by acid catalysis.

The following examples emphasise this property of cyclopropanols, but also indicate the stereospecificity of the proton, or deuteron uptake (de Puy and Breitbeil, 1963).

In acid solution, reaction is induced by protonation of the C–C bond and occurs with retention of the configuration at the reaction site. In base induced cleavage, on the other hand, reaction may be regarded as consequent on the development of carbanion character at the reaction site, and proton or deuteron addition then occurs with inversion of stereochemistry.

5.3. Nucleophilic and electrophilic fission of cyclopropanes. A cyclopropane carrying a group which stabilises a carbanion is subject to nucleophilic fission:

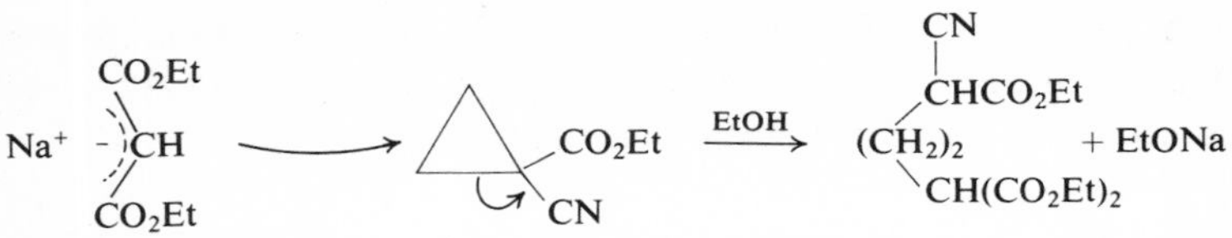

a reaction analogous with Michael addition, which emphasises the olefin-like conjugative character of the cyclopropyl residue.

Electrophilic fission of the cyclopropane ring follows the Markownikoff rule, e.g.:

However, the process is stepwise, and the following examples emphasise the importance of proton transfer within the initially formed cation (cf. Deno *et al.*, 1970)

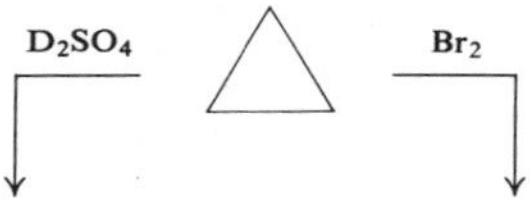

$$CH_2D.CH_2.CH_2.OSO_3D,$$
$$+CH_3.CHD.CH_2.OSO_3D,$$
$$+CH_3.CH_2.CHD.OSO_3D$$

$$CH_2Br.CH_2.CH_2Br,$$
$$+CH_3.CHBr.CH_2Br,$$
$$+CH_3.CH_2.CHBr_2$$

The course of these reactions may be represented:

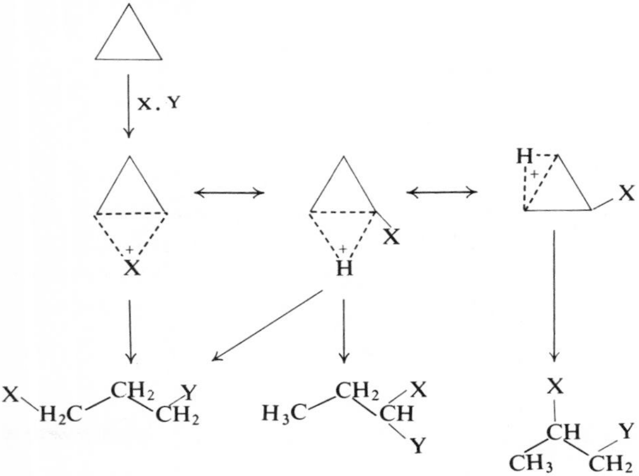

$(X.Y = Br.Br; D.SO_4D; RCO.Cl)$

i.e. by exchange of bonding involving X and ring hydrogens the charge introduced by the electrophilic addend X^+ becomes redistributed within the ring.

Although these reactions are represented using an edge protonated cyclopropane (A), the corner protonated alternative (B) may clearly be concerned in these rearrangements, cf. §8.11.

The ready interconversion of cyclopropyl methyl- and cyclobutyl-halides:

via rearrangement through a common carbonium ion has been discussed in chap. 3. A somewhat similar transformation which must, however, involve anionic rather than cationic carbon is observed in the reaction of a cyclopropyl methyl halide with lithium or magnesium (Kandil and Dessey, 1966). The metal derivative (14) is formed momentarily, but

CH_2-hal $\longrightarrow$ CH_2Mg-hal $\longrightarrow$

(14)

$CH_2{=}CHCH_2CH_2Mg$-hal

(15)

is transformed very rapidly to the butenyl derivative (15).

5.4. The cyclopropyl carbanion. Metal derivatives, or corresponding carbanions, in which the metal, or anionic change, is located directly on the cyclopropyl ring are, on the other hand, of some stability. The transformations:

Ph, Me $\xrightarrow{\text{BuLi}}$ Ph, Me $\xrightarrow{\text{MeOH}}$ Ph, Me

Ph′ Br   Ph′ Li   Ph′ H

$[\alpha]_D + 106°$ (16) $[\alpha]_D - 78°$

illustrate the formation and stereospecific substitution of a lithium cyclopropyl derivative (16). The steric stability of (16), i.e. its resistance to inversion, is attributable to the resistance to carbon–carbon bond angle widening which is involved in passing through the planar intermediate (17):

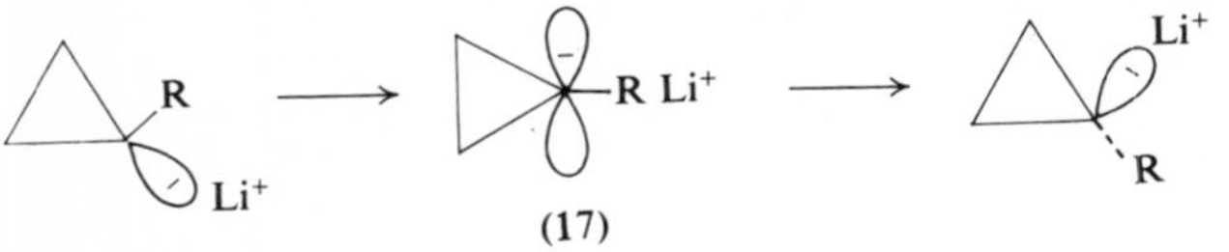

(17)

Similarly, steric stability of the cyclopropyl carbanion is implied in the case of the optically active nitrile (18), for which base catalysed hydrogen–deuterium exchange is found to be faster than racemisation (Walborsky and Motes, 1970).

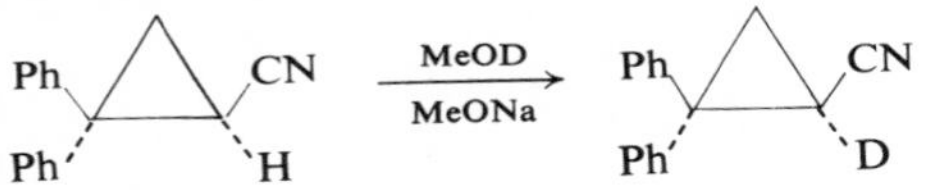

(18)

5.5. The stability of cyclopropanes. The cyclopropane ring is relatively rather stable to oxidation, e.g. carone (19)→*cis*-caronic acid (20).

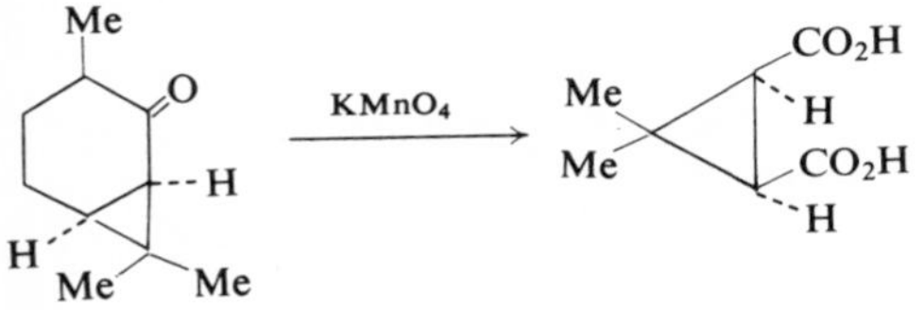

(19)   (20)

On strong heating, however, cyclopropanes may be cleaved and rearranged (cf. Frey, 1966):

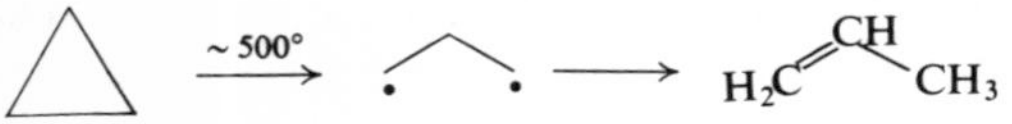

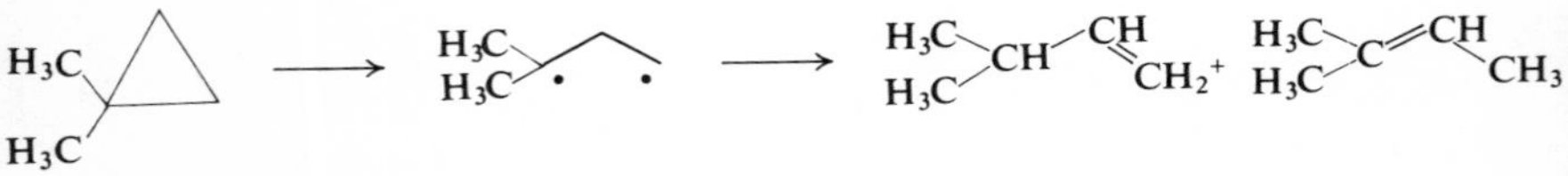

although the activation energy (*c.* 270 kJ/mole) which is required emphasises the thermal stability of the ring. In suitable cases formation of a more stable ring may occur rather than hydrogen transfer within the diradical, e.g.:

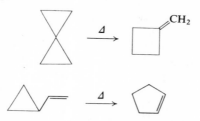

For vinyl cyclopropane rearrangement could be facilitated by allylic stabilisation in a radical intermediate (*A*), but a concerted 1,3-migration as in (*B*) must also be considered possible (Woodward and Hoffmann, 1970).

(A) (B)

On catalytic hydrogenation many cyclopropanes undergo hydrogenolysis, often rather slowly, but generally faster at palladium than at platinum. An electrophilic substituent, e.g. Ph, assists hydrogenolysis and is also found to influence the direction of ring fission (Irwin and McQuillin, 1968), e.g.:

R—△   (21)

|Pd/H$_2$

R = Ph   R = n-C$_6$H$_{13}$

Ph—H$_2$C—CH$_2$CH$_3$ n-C$_6$H$_{13}$—CH(CH$_3$)CH$_3$

5.6. Synthesis of cyclopropane and cyclopropane derivatives. Various routes to the synthesis of cyclopropane and its derivatives are outlined below:

(a) Also some rearrangement to γ-butyrolactone.
(b) The cyclopropanol is too alkali sensitive for conventional alkaline
    hydrolysis.

An interesting and valuable new method of cyclopropane synthesis is
based on hydroboration (Brown and Rhodes, 1969) of an allyl halide
followed by treatment with alkali, e.g.:

A similar sequence starting from propargyl bromide provides a convenient synthesis of cyclopropanol in good yield (Brown and Rhodes, 1969).

$$CH\equiv C-CH_2Br \xrightarrow{\;\gt BH\;} (\gt B)_2CHCH_2CH_2Br$$

$$\xrightarrow{NaOH} \;\gt B-\triangle \xrightarrow{H_2O_2} HO-\triangle$$

$\gt BH = 9$-borabicyclo[3,3,1]nonane, obtained from cyclo-octa-1,5-diene and diborane.

$$\text{(cyclo-octa-1,5-diene)} \xrightarrow{B_2H_6} \text{(BH)}$$

Homoallylic displacement is the basis of a useful cyclopropane synthesis, e.g.:

or:

$$BrCH_2-CH\!=\!CH-CH_2Br \xrightarrow[\text{NaOEt}]{CH_2(CO_2Et)_2}$$

The following examples make use of benzyl reactivity (Baumgardner, 1961).

(NBS = N-Bromosuccinimide)

5.7. Cyclopropane syntheses by carbene addition. Addition of a carbene to an olefin provides a valuable route for the synthesis of cyclopropanes. Halogenocarbenes in particular are very conveniently generated *in situ* by a variety of methods, e.g.:

$$t\text{-BuO}^- + CHCl_3 \rightarrow t\text{-BuOH} + \overset{-}{C}Cl_3; \ \overset{-}{C}Cl_3 \rightarrow :CCl_2 + Cl^-,$$

$$n\text{-BuLi} + CH_2Cl_2 \rightarrow C_4H_{10} + LiCHCl_2 \rightarrow :CHCl + LiCl,$$

$$Br_3C.CO_2Na \overset{\triangle}{\rightarrow} \overset{-}{C}Br_3 + CO_2 + Na^+ \rightarrow :CBr_2 + NaBr,$$

$$PhHgCBr_3 \overset{\triangle}{\rightarrow} PhHgBr + :CBr_2.$$

Dihalogenocarbenes are found to add stereospecifically, e.g.:

although a monohalogenocarbene necessarily leads to two isomers, e.g.:

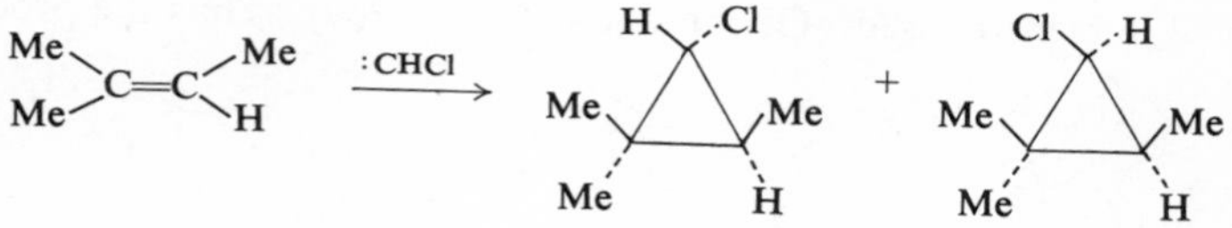

For halogenocarbenes formed by use of a lithium alkyl:

$$BuLi + HCBr_3 \rightarrow C_4H_{10} + LiCBr_3$$

the intermediate lithium derivative rather than the free carbene may be the addend.

A reaction formally similar to carbene addition, namely base catalysed reaction between ethyl chloroacetate and an $\alpha\beta$-unsaturated ester, may be formulated as Michael addition followed by ring closure.

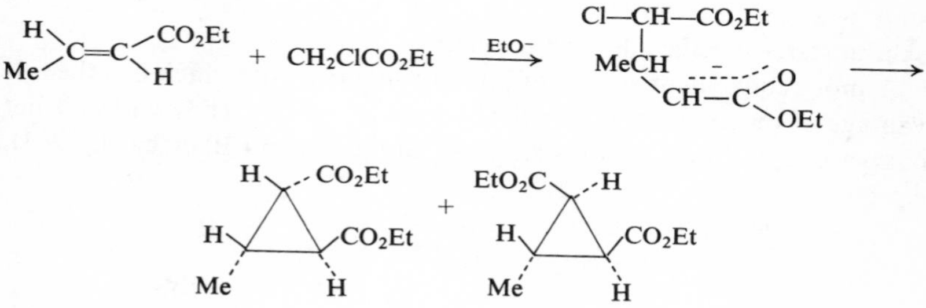

Methylene, $:CH_2$, generated by photolysis of diazomethane: $CH_2N_2 \xrightarrow{h\nu}$ $:CH_2 + N_2$, adds mainly stereospecifically, but insertion into C–H bonds constitutes an unwanted side reaction:

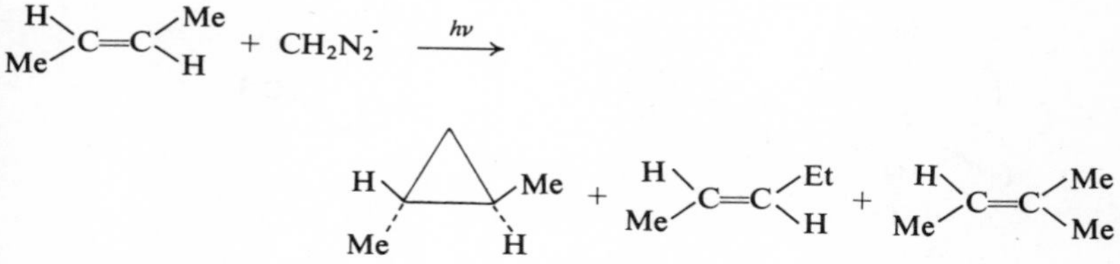

For methylene, $:CH_2$, the ground state is the triplet or diradical (A, R = H), whilst the excited level is the singlet state (B, R = H). Spin pairing in the singlet state permits addition to an olefin in a concerted

stereospecific process which is symmetry allowed, whilst the triplet
state may lead to non-stereospecific addition through rotation in the
diradical intermediate (cf. Gilchrist and Rees, 1969). Thus the product

ratios depend upon the mode of preparation and extent of singlet→
triplet deactivation of methylene. For the dihalogenocarbenes the ground
state is a singlet as in (B, R = hal.) and dihalogenocarbenes therefore
exhibit stereospecific addition to olefins.

A more convenient source of a methylene reagent, which has the ad-
vantage of reacting stereospecifically, makes use of CH_2I_2 with a zinc/
copper couple (the Simmons reaction, Simmons and Blanchard, 1964),
e.g.:

A third example:

(22)

illustrates stereospecific entry of the methylene residue attributed to co-ordination of the Zn-reagent to a proximate hydroxyl group, which was made use of in the synthesis of the sesquiterpene thujopsene (22) (Dauben and Ashcraft, 1963).

A carbene unit may also be generated by decomposition of a diazo-alkane catalysed by a cuprous halide or by copper bronze; this gives different product ratios from photo-decomposition.

$$CH_2{=}C(OEt)_2 + RCHN_2 \xrightarrow{Cu_2Br_2} R \triangle \begin{smallmatrix} OEt \\ OEt \end{smallmatrix}$$

$$CH_2{=}CHOEt + N_2CHCO_2Et$$

EtO△H / H CO₂Et   +   EtO△CO₂Et / H H

	63%	11%
Cu	63%	11%
hv	31%	16%

Addition of, e.g., diazomethane to an activated olefin followed by thermal or photochemical decomposition of the intermediate pyrazoline also offers a useful method of cyclopropane synthesis.

EtO₂C, Pr^i >C=C< H, CO₂Et $\xrightarrow{CH_2N_2}$ (pyrazoline: N=N–CH₂ ring, EtO₂C, Pr^i / H, CO₂Et) $\xrightarrow{-N_2}$

EtO₂C△H / Pr^i CO₂Et   +   EtO₂C△CO₂Et / Pr^i H

A pyrazoline may also be generated:

H, Ph >C=C< COPh, H $\xrightarrow{N_2H_4}$ (HN–N ring =Ph, Ph) $\xrightarrow{-N_2}$

Ph△H / H Ph   +   Ph△Ph / H H

An internal carbene insertion reaction, e.g.:

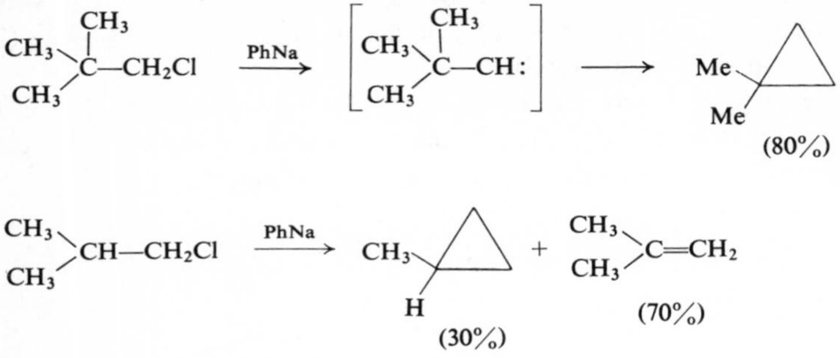

(80%)

(30%)   (70%)

may also lead to a cyclopropane synthesis in suitable cases (Friedman and Berger, 1961).

5.8. Cyclopropene. The olefinic bond in cyclopropene (23) (Closs, 1966) is unusually short, and the olefinic ν_{CH} [3080–3100 cm^{-1}] is correspondingly high, cf. chap. 1. In consequence the olefinic hydrogen

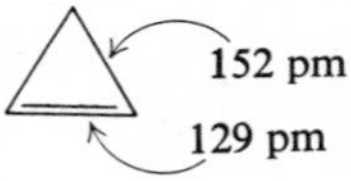

152 pm

129 pm

(23)

displays the acidity to be expected of a C–H bond of high *s*-character:

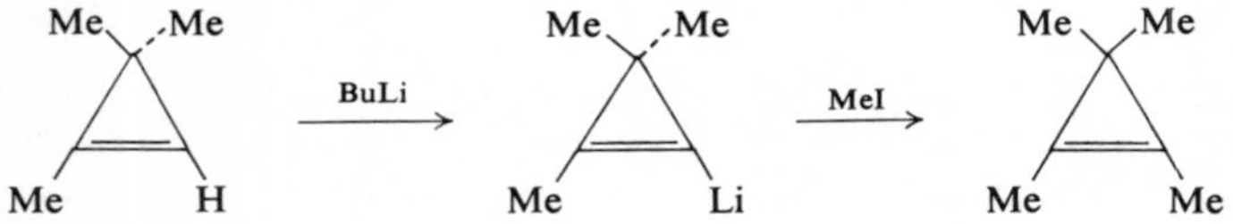

and the strained olefinic bond is also active in the Diels–Alder addition:

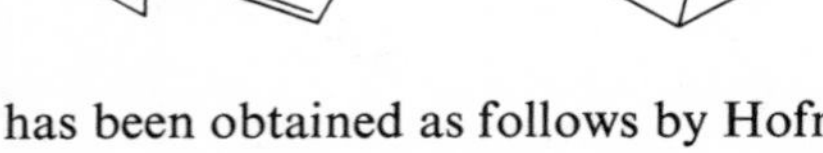

Cyclopropene has been obtained as follows by Hofmann elimination.

300°

(Pt catalyst)

The following cyclopropene synthesis makes use of the principle of internal carbene addition:

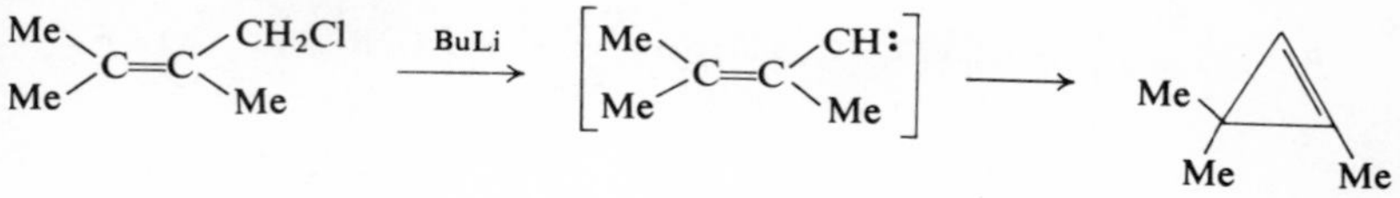

5.9. The cyclopropenium ion. This ion (24) (Breslow, 1970), represents the simplest aromatic ring structure of $(4n + 2)$ π-electrons, $n = 0$, in terms of Hückel's definition. Various cyclopropenium salts have been

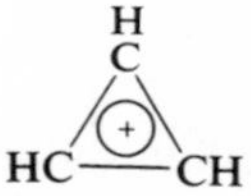

(24)

synthesised by addition of a suitable carbene to an acetylene. Diphenyl acetylene and phenyl chlorocarbene, generated from benzylidene chloride, give triphenyl chlorocyclopropene which is converted into triphenyl cyclopropenium chloride as shown below.

$$PhCHCl_2 + t\text{-}BuOK \longrightarrow PhCCl \; ; \; PhC{\equiv}CPh + PhCCl \longrightarrow$$

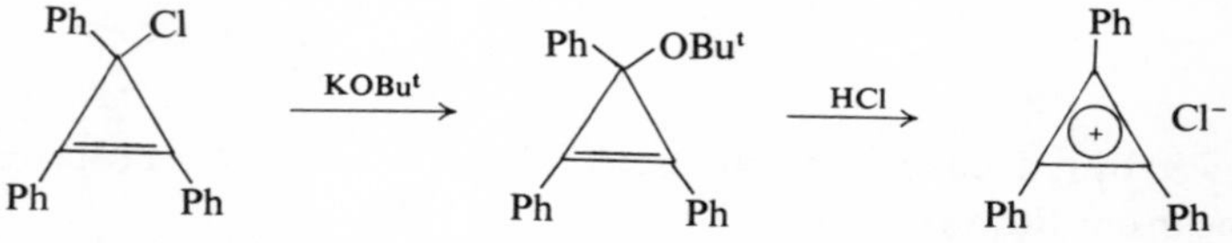

In a second example, addition of ethoxycarbonyl carbene, obtained by copper catalysed decomposition of ethyl diazoacetate, to oct-4-yne gives ethoxycarbonyl di-n-propyl cyclopropene which is converted into the cyclopropenium salt:

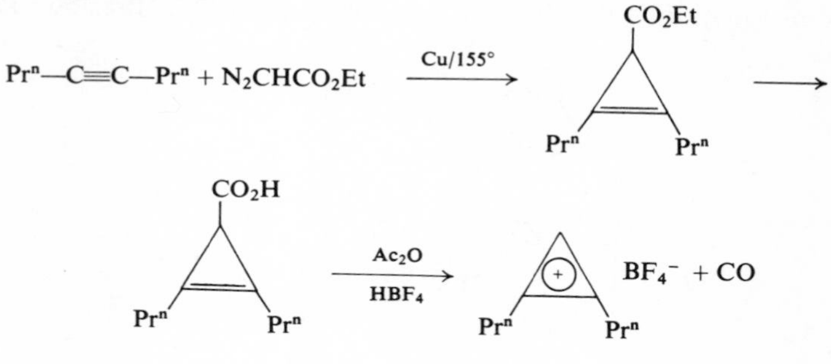

The decarboxylation procedure which depends on loss of carbon monoxide from an acylonium ion is of interest, viz.:

$$RCO_2H \xrightarrow{Ac_2O} R.CO.OC.OCH_3 \xrightarrow{HBF_4} R\overset{+}{C}O\ BF_4^-;\ R\overset{+}{C}O\ BF_4^-$$

$$R^+\ BF_4^- + CO,$$

where R = cyclopropenyl.

Salts of the parent unsubstituted cyclopropenium ion (24) have been obtained:

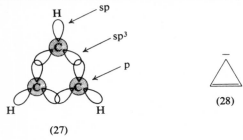

(25)

The $SbCl_6^-$ salt forms a white solid, stable for some time at the ordinary temperature and for long periods at $-20°$.

The chlorocyclopropene (25) used in this preparation was obtained by heating sodium trichloroacetate in presence of trichloroethylene

$$CCl_3CO_2Na \longrightarrow :CCl_2 + CO_2 + NaCl;\ :CCl_2 +$$

(26)

followed by reductive dechlorination of the tetrachlorocyclopropene (26) by means of Bu_3SnH.

The cyclopropenium ion shows ν_{C-H} at 3105 cm^{-1} in the infrared, and an n.m.r. proton signal at $\tau -1.1$. However, most interesting is the high value of the ^{13}C-proton coupling constant, $J_{13_{C-H}}$, of 265 Hz, which corresponds with some 53 per cent of s-character in the C–H bond. This information is rationalised in a formulation (27) (Breslow, 1970)

(27)

(28)

in which the sp^3 lobes join the carbon framework and the shaded areas represent the p-lobes which overlap above and below the plane of the ring to constitute the delocalised π-system.

5.10. The cyclopropenyl anion. The cyclopropenyl anion (28) differs from the cyclopropenium cation (24) in that the latter conforms to Hückel's $4n + 2$ π-electron criterion of aromaticity, whilst the anion (28) represents a $4n$ π-electron system ($n = 1$) which is regarded as 'anti-aromatic' and relatively destabilised (Breslow, 1968). For this reason the CH_2 group of a cyclopropene is a weaker carbon acid than the CH_2 group of a similar cyclopropane.

5.11. Cyclopropanones. A number of cyclopropanones (29) (Turro, 1969) have been prepared by reaction of diazomethane with excess of a ketene:

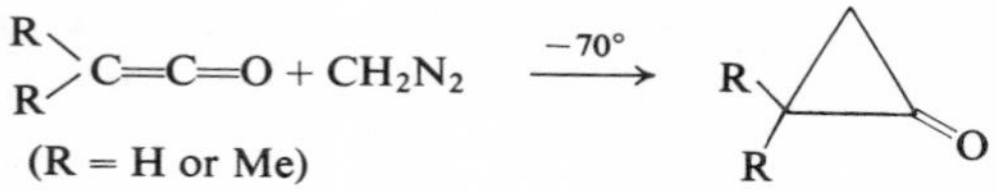

(29)

Excess diazomethane leads to formation of cyclobutanones through further reaction of the cyclopropanone which is first formed.

Cyclopropanones show the expected high carbonyl stretching frequency, $\nu_{C=O}$ 1813 cm^{-1}. The $C=O$ bond length is somewhat shorter (118 pm) and the CH_2—CH_2 bond length (158 pm) is somewhat longer than the normal.

The following reactions illustrate the reactivity of the carbonyl group towards addition:

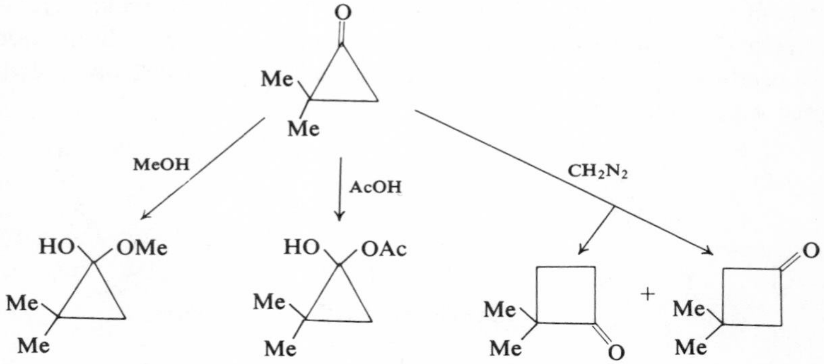

Cyclopropanones, however, also exhibit cycloaddition reactions of the type:

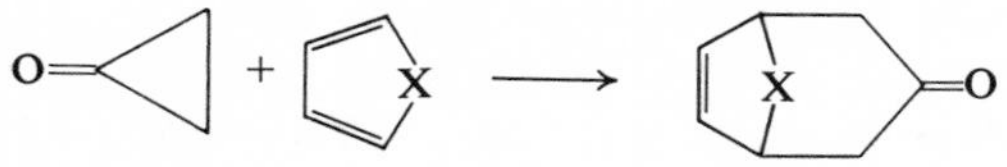

This is attributable to a relatively easy cyclopropyl → allyl transformation promoted by the strong carbonyl and weakened CH_2—CH_2

(30)

bonds noted above. The allyl cationic form (30) then constitutes a two π-electron addend for cyclo-addition:

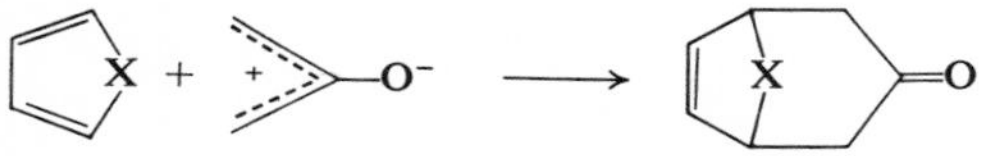

5.12. Cyclopropenones. These compounds (31), (Krebs 1965; Breslow, 1967), have been synthesised by addition of dichlorocarbene to an acetylene followed by hydrolysis of the intermediate dichlorocyclopropene.

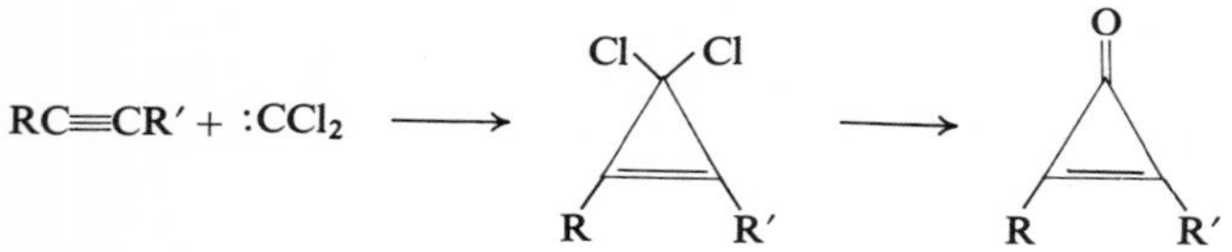

(31)

Cyclopropenone itself (32) was obtained from the tetrachlorocyclopropene (26) noted above, which was reduced by tri-n-butyl tin hydride to a product containing the dichlorocyclopropene (33) which was hydrolysed with water:

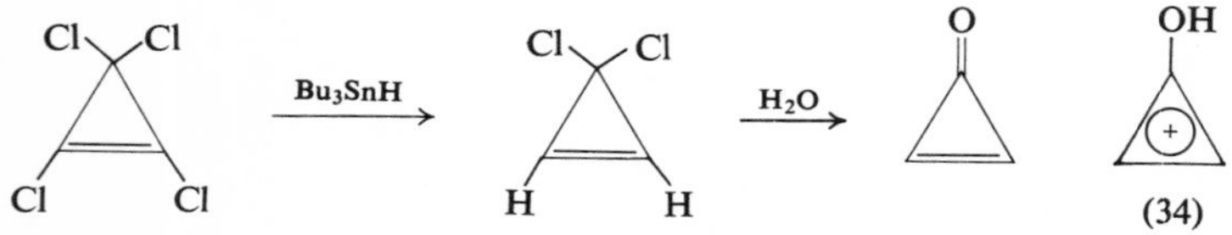

(26)   (and other products)   (32)

(33)

Cyclopropenones show two infrared bands in the 1800 cm^{-1} region, e.g. for cyclopropenone (32) at 1835 and 1870 cm^{-1}. The carbonyl group does not appear to show the addition reactions which are characteristic of cyclopropanone, but in acid solution it is protonated to give the cyclopropenium form (34), which shows no infrared carbonyl absorption.

The stability of the cyclopropenone ring system is ascribed (Clark, 1970) to the fact that although decomposition to acetylene and carbon monoxide is thermodynamically favourable, it is also a symmetry forbidden process which raises the activation energy required (cf. chap. 4).

5.13. Naturally occurring cyclopropanes and cyclopropenes. Mention should also be made of the natural occurrence of cyclopropane and cyclopropene derivatives, e.g. lactobacillic acid (35) from *lactobacillus arabinosus* and sterculic acid (36) from the oil of *sterculia foetida*.

$$CH_3(CH_2)_5 \underset{(35)}{\overline{\bigtriangleup}} (CH_2)_9CO_2H \qquad CH_3(CH_2)_7 \underset{(36)}{\overline{\bigtriangleup}} (CH_2)_7CO_2H$$

The recent synthesis of sterculic acid illustrates a number of points of interest. Direct addition of methylene, generated from diazomethane, to stearolic acid failed, but use of ethyl diazoacetate as a source of ethoxy-carbonyl carbene led to the successful synthesis below (Gensler *et al.*, 1970).

$$CH_3.(CH_2)_7C{\equiv}C.(CH_2)_7.CO_2Me + N_2CHCO_2Et \xrightarrow{Cu}$$

$$CH_3.(CH_2)_7.\underset{\underset{CH.CO_2Et}{\diagdown\diagup}}{C{=}C}.(CH_2)_7.CO_2Me \xrightarrow{KOH \ aq.}$$

$$CH_3.(CH_2)_7.\underset{\underset{CH.CO_2H}{\diagdown\diagup}}{C{=}C}.(CH_2)_7.CO_2H \xrightarrow{(COCl)_2}$$

$$CH_3.(CH_2)_7.\underset{\underset{CH.COCl}{\diagdown\diagup}}{C{=}C}.(CH_2)_7.COCl \xrightarrow{ZnCl_2}$$

$$CH_3.(CH_2)_7.\underset{\underset{CH}{\diagdown\underset{\oplus}{}\diagup}}{C{-}C}.(CH_2)_7.COCl \xrightarrow{MeOH}$$

$$CH_3.(CH_2)_7.\underset{\underset{CH}{\diagdown\underset{\oplus}{}\diagup}}{C{=}C}.(CH_2)_7.CO_2Me \xrightarrow{NaBH_4}$$

$$CH_3.(CH_2)_7.\underset{\underset{CH_2}{\diagdown\diagup}}{C{=}C}.(CH_2)_7.CO_2Me$$

The intermediate cyclopropene carboxylic acid was converted to the cyclopropenium salt by decarbonylation of the acid chloride:

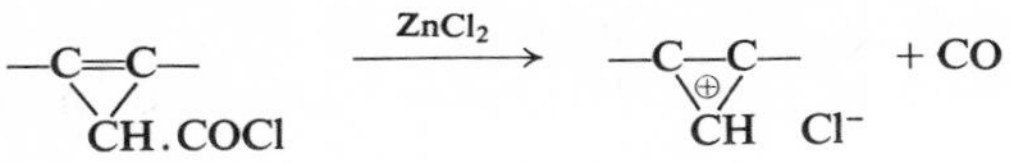

5.14. Cyclobutanes

Thermal decomposition. Cyclobutanes are resistant to thermolysis (cf. chap. 4), although cyclobutane may be decomposed to ethylene at a temperature of 500°.

Thermolysis is assisted by a radical stabilising group as in vinyl cyclobutane.

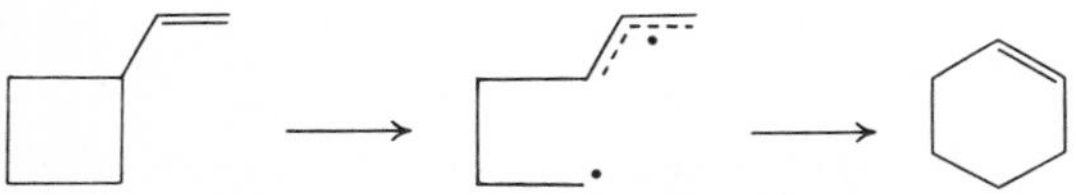

The pyrolysis of α-pinene (37), which is an industrially important process, also depends on thermal fission of a vinyl cyclobutane system to give an intermediate allyl radical (cf. Banthorpe and Whittaker, 1966).

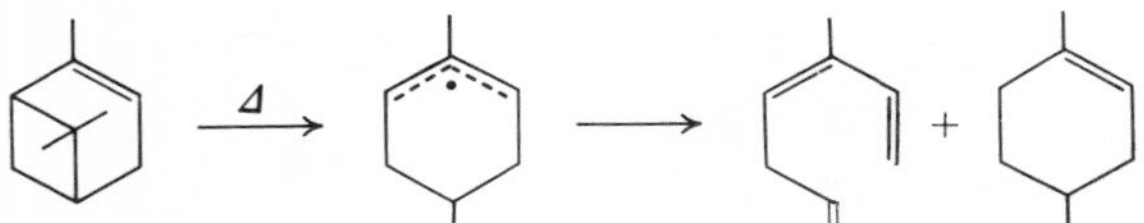

Displacement reactions. In displacement reactions of cyclobutyl derivatives rearrangement is frequently observed. Thus conversion of cyclobutanol to the chloride leads also to cyclopropyl methyl chloride and but-3-enyl chloride (cf. chap. 3).

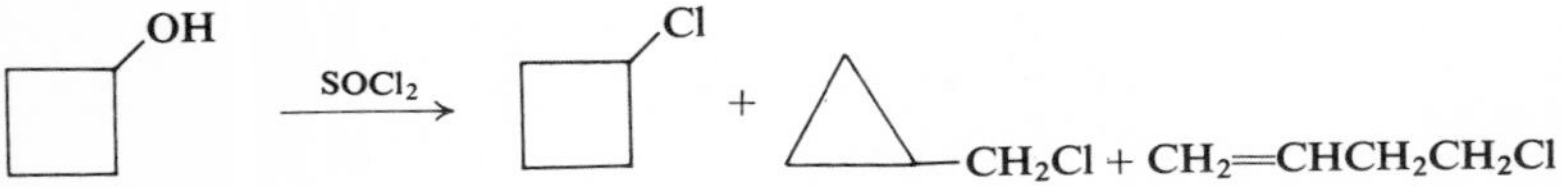

Collapse to a cyclopropane is illustrated also by the facile Favorskii rearrangement of 2-bromocyclobutanone.

Ring enlargement is also observed, as in the Wagner–Meerwein rearrangement of cyclobutylcarbinol on treatment with hydrogen bromide.

A cyclobutyl halide is found to undergo rearrangement to the but-3-enyl derivative on attempted formation of the cyclobutyl Grignard derivative (Kandil and Dessey, 1966).

Cyclobutyl halides or the toluene-*p*-sulphonate (38) are solvolysed relatively rapidly, at about the same rate as the corresponding allyl derivative. The high rates, and the formation of a mixed product

(38)   (39)

are attributable to charge delocalisation in an intermediate non-classical cation such as (39) (cf. §3.3).

Stability to oxidation. The stability of the cyclobutane ring to oxidation is indicated by the oxidative degradation of α-pinene (40) to pinonic

(40)   (41)   (42)

acid (41) and thence to *cis*-norpinic acid (42). The cyclobutane ring is stable to acid since *cis*- (43) and *trans*-norpinic acid (44) may be inter-converted by vigorous acid treatment:

(43)   (44)

Pinonic acid (41) on acid treatment undergoes ring fission but this de-

(41)

pends on the ease of protonation of the keto group, and on the stabil-isation of a cationic charge at the tertiary $-CMe_2$ group.

5.15. Cyclobutanone. Cyclobutanone (45) may be reduced to cyclo-butanol (46) in high yield, but owing to the very ready cyclobutyl →

(45)   (46)

cyclopropylmethyl interconversion (cf. chap. 3) attempted re-oxidation of cyclobutanol under acid conditions is unsatisfactory.

Cyclobutanone may be obtained in useful yield by the sequence:

starting from pentaerythritol bromide (Roberts and Sauer, 1949), or from ketene and diazomethane (Semenov *et al.*, 1956):

$$CH_2{=}C{=}O + \text{excess } CH_2N_2 \longrightarrow$$

5.16. Cyclobutene. This compound (47) which is prepared in good yield by the Hofmann, or the Cope elimination reaction (Roberts and Sauer, 1949) is much less thermally stable than cyclobutane, and on heating

(47)

cyclobutene is readily transformed to buta-1,3-diene. This type of thermal reaction is typical of cyclobutenes, and is also an important instance of the group of stereospecific electrocyclic reactions which are discussed

more fully in chap. 4.

5.17. Synthesis of cyclobutanes.   Various cyclobutane syntheses are outlined below:

1,3-Substituted cyclobutanes have been obtained in a similar manner:

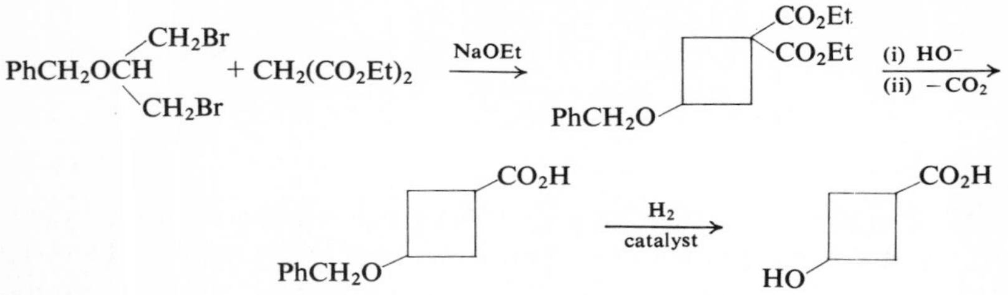

Dimerisation of highly reactive olefins is a valuable general method of synthesis. Dimerisation of dimethyl ketene is spontaneous and the

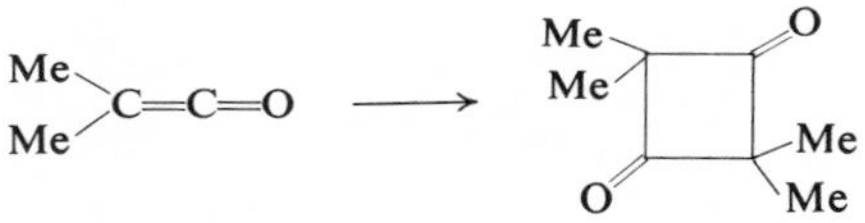

ease of reaction is accounted for by the vinylinium ylid character of a ketene: $>C=C^+—O^-$. Dimerisation is regarded as a concerted process and the orientation of the product is rationalised as follows (Huisgen and Otto, 1968).

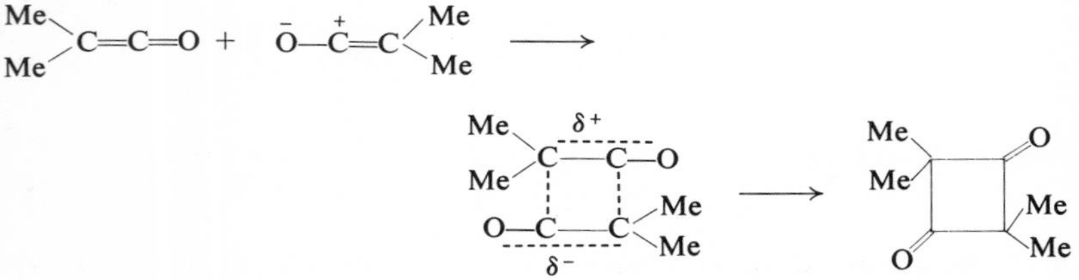

However, the orientation in this reaction is exceptional; more generally dimerisation leads mainly to the 1,2-substituted cyclobutanes (Roberts, 1962). Allene, for example, gives mainly 1,2-dimethylene cyclobutane:

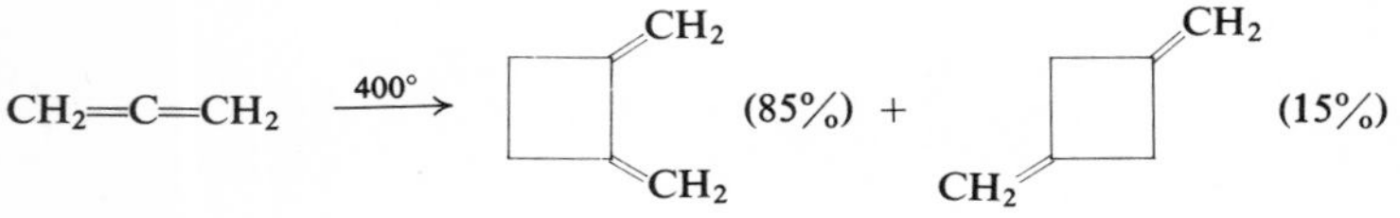

Acrylonitrile similarly gives *cis*- and *trans*-1,2-dicyanocyclobutanes:

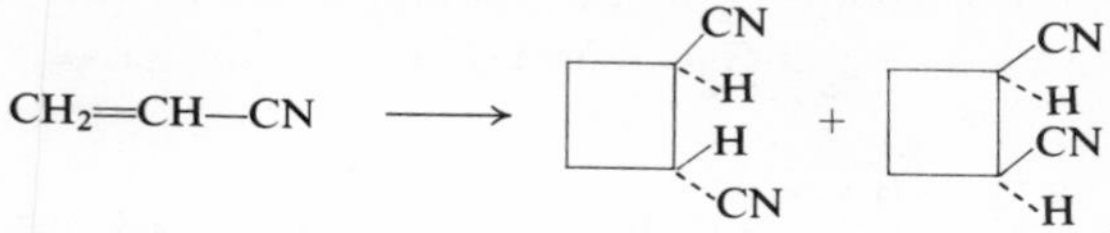

Polyfluoroethylenes, in particular, dimerise rather readily on heating. The following example, starting from difluorodichloroethylene illustrates this type of synthesis, and its value as a route to the interesting substance 1,2-dihydroxycyclobut-1-en-3,4-dione (48). This substance behaves as a strong acid (pK_2 2.2) since the dianion (49) corresponds with an aromatic type structure (50) of six π-electrons (cf. Maahs and Hegenberg, 1966).

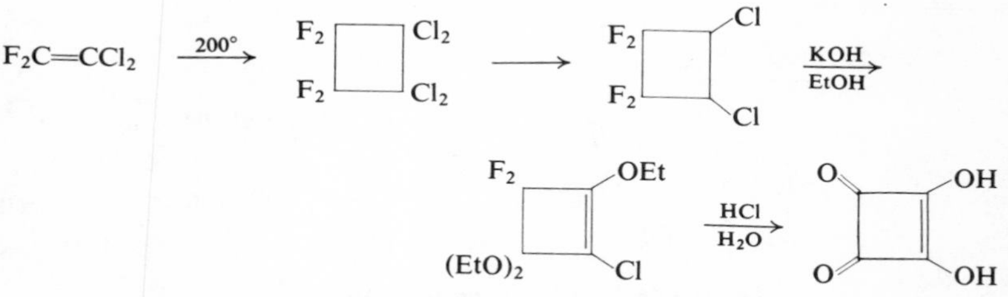

(48)

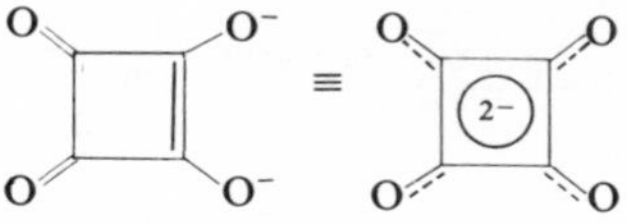

(49)   (50)

Cyclobutanes are also obtained by photoaddition of suitable olefins (cf. chap. 4). The following example illustrates the use of this method as the first step in Corey's synthesis of caryophyllene (Corey, 1964).

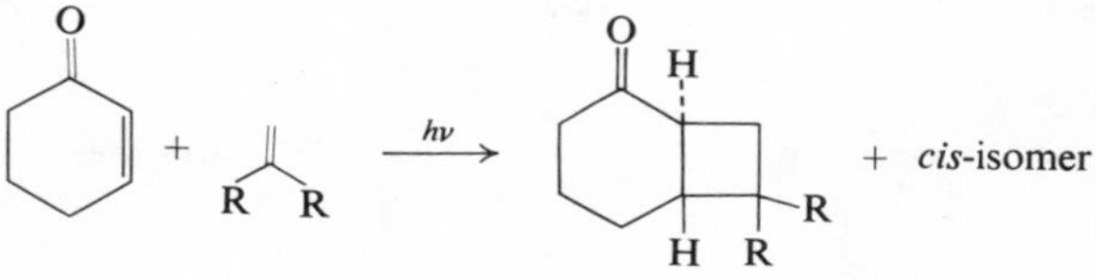

(51)

5.18. Cyclobutadiene. This highly strained and reactive structure (54) is stabilised by co-ordination in a range of transition metal complexes

(Cava and Mitchell, 1967). The free parent cyclobutadiene was obtained from the iron tricarbonyl complex (52) prepared by Pettit (1965, 1966, 1967), cf. Rosenblum and Gatronis (1967).

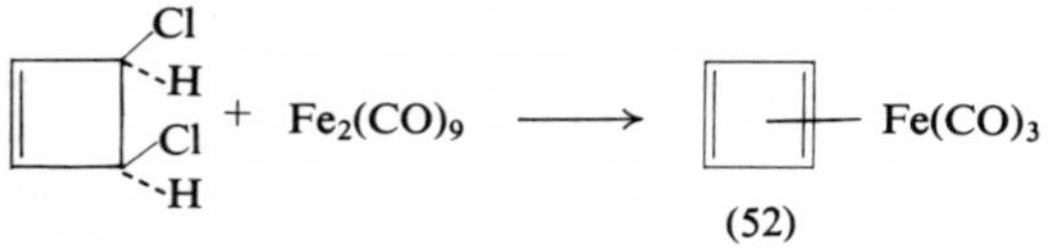

(52)

The required dichlorocyclobutene (53), was obtained from cyclo-

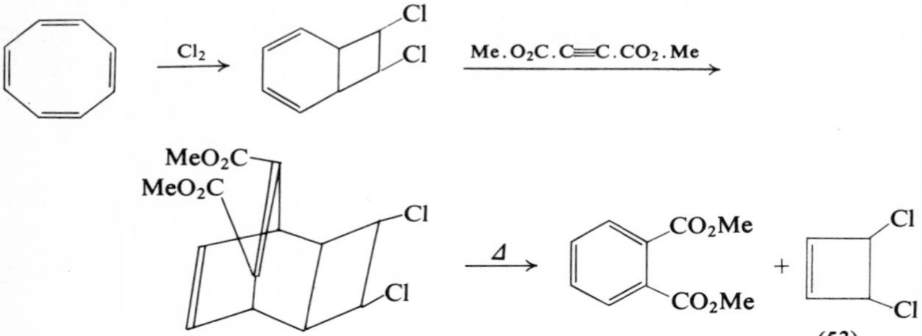

octatetraene as a mixture of *cis-* and *trans-* isomers.†

Oxidation of the iron tricarbonyl complex by means of ceric ammonium

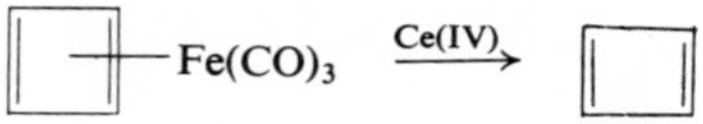

(54)

nitrate releases cyclobutadiene as a short-lived product, which rapidly dimerises, but which may be trapped as an adduct, e.g. (55) with phenyl acetylene, (56) with methyl propiolate, or with other acetylenes (Pettit, 1966):

(54)

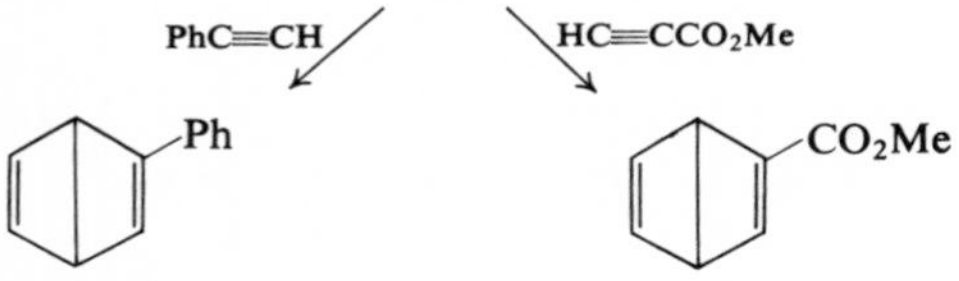

PhC≡CH   HC≡CCO₂Me

(55)   (56)

† Reactions of cyclo-octatetraene *via* a bicyclo[4,2,0]octatriene are discussed in §9.5.

These reactions also represent an interesting route to the synthesis of derivatives of bicyclo[2,2,0]hexadiene ('Dewar benzene') which are discussed more fully in chap. 8.

On theoretical grounds cyclobutadiene is recognised as having the rectangular diene structure (54) rather than the square diradical formulation (57) (Dewar and Gilcher, 1965). In agreement, the adducts with dimethyl maleate (58) and dimethyl fumarate (59) are formed stereospecifically, i.e. without evidence of stereomutation in a radical intermediate.

(57)   (58)   (59)

The geometry of cyclobutadiene is ideal for $4 + 2$ Diels–Alder cycloaddition, as in the formation of the dimers (60) and (61), the *syn*

(60)   (61)

isomer (60) being the major product, formed by *endo*-addition. In the presence of silver fluoroborate the *anti*-isomer (61) is rapidly transformed to cyclo-octatetraene (62) (Pettit, 1967), which also illustrates

(61)   (62)

the manner in which metal ion complexing may in suitable cases facilitate a reaction; the transformation (61)→(62) requires a process of disrotation, cf. chap. 4. The extra electrons and orbitals available in the metal ion make possible what is otherwise a symmetry disallowed process. However, both the *syn* and *anti*-isomers (60) and (61) are transformed to cyclo-octatetraene on heating, but much more slowly.

References

Banthorpe, D. V. and Whittaker, D. (1966). *Quarterly Reviews*, **20**, 373.

Baumgardner, C. L. (1961). *J. Amer. Chem. Soc.*, **83**, 4420.

Breslow, R. (1967). *J. Amer. Chem. Soc.*, **89**, 3073; (1968). *Chemistry in Britain*, **4**, 100; (1970). *J. Amer. Chem. Soc.*, **92**, 984.

Brown, H. C. and Rhodes, S. P. (1969). *J. Amer. Chem. Soc.*, **91**, 2149, 4306.

Cava, M. P. and Mitchell, M. J. (1967). *Cyclobutadiene and related compounds*, Academic Press.

Clark, D. T. (1970). *Chemical Communications*, 147.

Closs, G. L. (1966). *Advances in Alicyclic Chemistry*, **1**, 53.

Corey, E. J. (1964). *J. Amer. Chem. Soc.*, **86**, 485, 5570.

Dauben, W. G. and Ashcraft, A. C. (1963). *J. Amer. Chem. Soc.*, **85**, 3673.

Deno, N. C., Billups, W. E., LaVietes, D., Scholl, P. E. and Schneider, S. (1970). *J. Amer. Chem. Soc.*, **92**, 3700.

Dewar, M. J. S. and Glicher, G. J. (1965). *J. Amer. Chem. Soc.*, **87**, 3255.

Friedman, L. and Berger, J. G. (1961). *J. Amer. Chem. Soc.*, **83**, 493, 500.

Frey, H. M. (1966). *Advances in Physical Organic Chemistry*, **4**, 147.

Gensler, W. J., Floyd, M. B. and Yanase, R. and Rober, K. W. (1970). *J. Amer. Chem. Soc.*, **92**, 2472.

Gilchrist, T. L. and Rees, C. W. (1969). *Carbenes, Nitrenes and Arynes*, Nelson.

Huisgen, R. and Otto, P. (1968). *J. Amer. Chem. Soc.*, **90**, 5302.

Irwin, W. J. and McQuillin, F. J. (1968). *Tetrahedron Letters*, 2195.

Kandil, S. A. and Dessey, K. E. (1966). *J. Amer. Chem. Soc.*, **88**, 3027; cf. Patel, D. J., Hamilton, C. L., and Roberts, J. D. (1965). *ibid.*, **87**, 5144.

Krebs, A. W. (1965). *Angewandte Chemie* (Internat. Edn.), **4**, 10.

Maahs, G. and Hegenberg, P. (1966). *Angewandte Chemie* (Internat. Edn.), **5**, 888.

Pettit, R. (1965). *J. Amer. Chem. Soc.*, **87**, 3253, 3255; (1966). *ibid.*, **88**, 623; (1967). *ibid.*, **89**, 4788.

de Puy, C. H. and Breitbeil, F. W. (1963). *J. Amer. Chem. Soc.*, **85**, 2176.

Roberts, J. D. (1962). *Organic Reactions*, **12**, 1.

Roberts, J. D. and Sauer, C. W. (1949). *J. Amer. Chem. Soc.*, **71**, 3925.

Rosenblum, M. and Gatronis, C. (1967). *J. Amer. Chem. Soc.*, **89**, 5074.

Schleyer, P. von R., Dine, G. W. van, Schöllkopf, U. and Paust, J. (1966). *J. Amer. Chem. Soc.*, **88**, 2868.

Semenov, D. A., Cox, E. F. and Roberts, J. D. (1956). *J. Amer. Chem. Soc.*, **78**, 3221.

Simmons, H. E. and Blanchard, E. P. (1964). *J. Amer. Chem. Soc.*, **86**, 1337, 1347.

Turro, N. J. (1969). *Accounts of Chemical Research*, **2**, 25.

Walborsky, H. M. and Motes, J. M. (1970). *J. Amer. Chem. Soc.*, **92**, 2445, 3697.

Woodward, R. B. and Hoffmann, R. (1970). *The Conservation of Orbital Symmetry*, Verlag Chemie/Academic Press.

6 Cyclopentane, cyclohexane and cycloheptane derivatives

6.1. Cyclopentanes. Syntheses of various cyclopentane derivatives starting from adipic acid or from cyclopentadiene are outlined below.

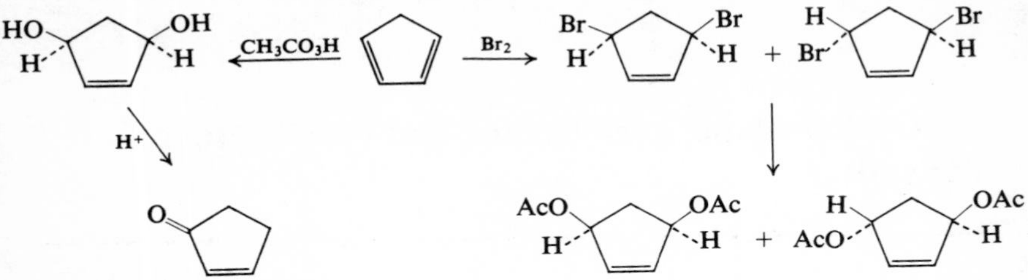

The condensation:

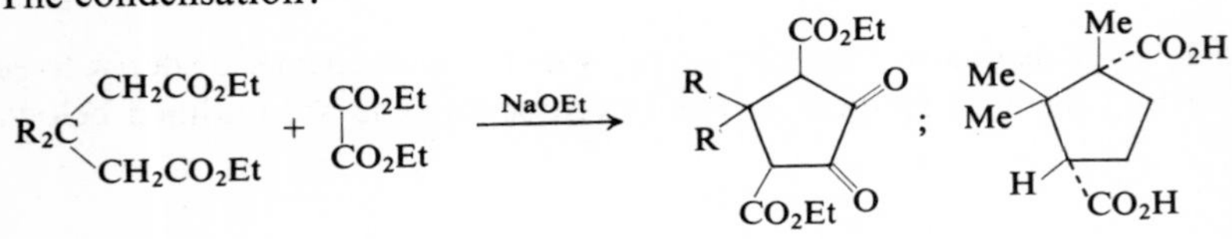

(1)

in the case $R_2 = Me_2$, was the basis of the synthesis of ($\pm$)-camphoric acid (1) (Komppa, 1903, 1909).

6.2. Cyclohexanes. Cyclohexane derivatives are often available by catalytic hydrogenation of an aromatic substance, e.g.:

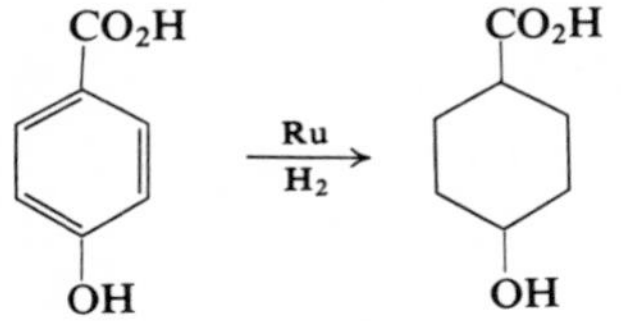

(Ruthenium is preferred as catalyst in this case since hydrogenation at platinum is accompanied by hydrogenolysis of the hydroxyl group.) Hydrogenation of a phenol in alkaline solution permits direct hydrogenation to the cyclohexanone. In this way resorcinol gives cyclohexane-1,3-dione (2), also known as dihydroresorcinol.

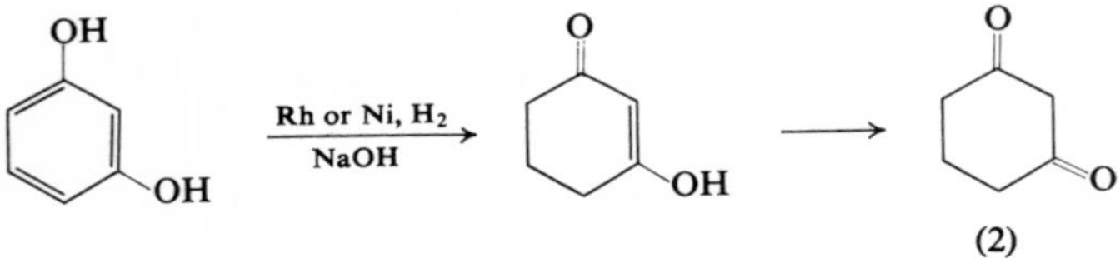

(2)

Birch reduction of an aromatic ring, using sodium or lithium in liquid ammonia with an alcohol, normally leads to a 1,4-dihydro benzene derivative, the orientation of hydrogen addition depending on the nature of the substituents, i.e. 1,4-addition as in (3) when R is an electron withdrawing group, e.g. CO_2H, and 2,5-addition as in (4) when R is electron donating, e.g. MeO, Me (Birch, 1950, 1958; cf. House, 1965).

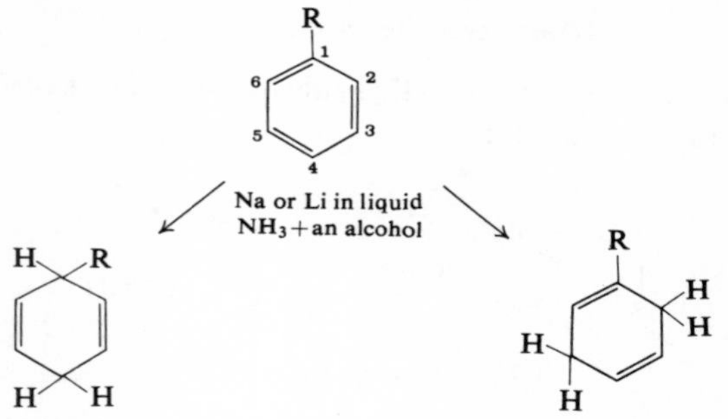

(3)   (4)

The Birch method is particularly valuable when applied to a phenol methyl ether which gives a vinyl ether, which in turn may be hydrolysed by acid to the ketone, e.g.:

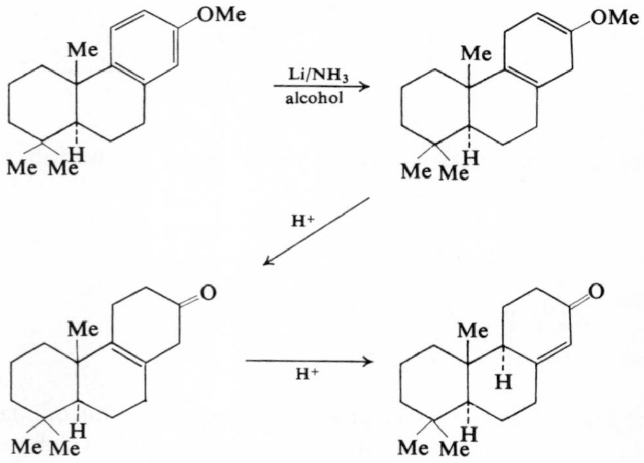

Cyclohexanes are available also by a number of methods based on cyclisation. Cyclisation may be brought about by generating an electrophilic centre in 1,6-relation to an olefinic bond (Johnson *et al.*, 1964), e.g.:

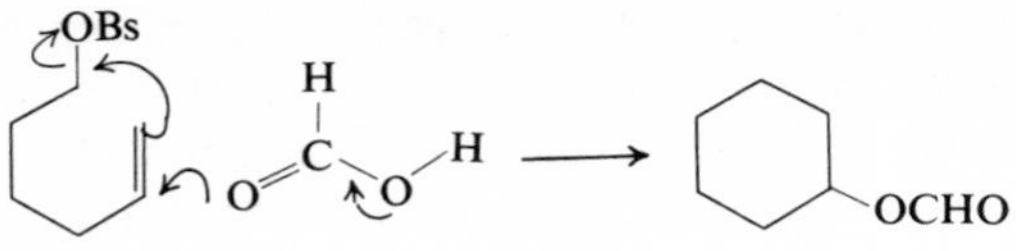

(OBs = bromobenzene-*p*-sulphonyl)

It may be noted, however, that alkyl substitution on the olefinic bond is important. In the example:

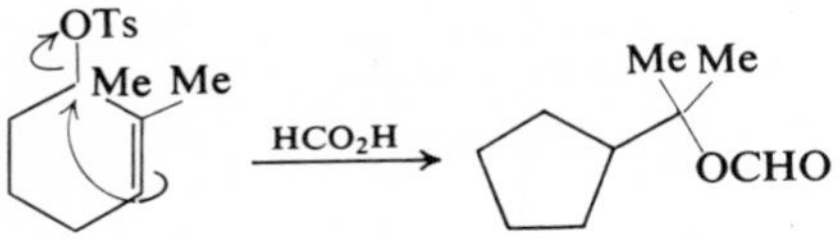

(OTs = toluene-*p*-sulphonyl)

Markownikoff addition necessarily leads to a cyclopentane derivative (Johnson and Owyang, 1964). The cyclisation of (−)-linalool (5) to give (+)-α-terpinyl acetate (6), amongst other products

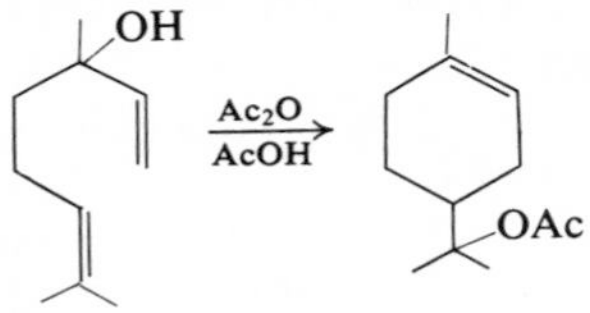

(5)   (6)

illustrates the concerted nature of this type of cyclisation, viz.:

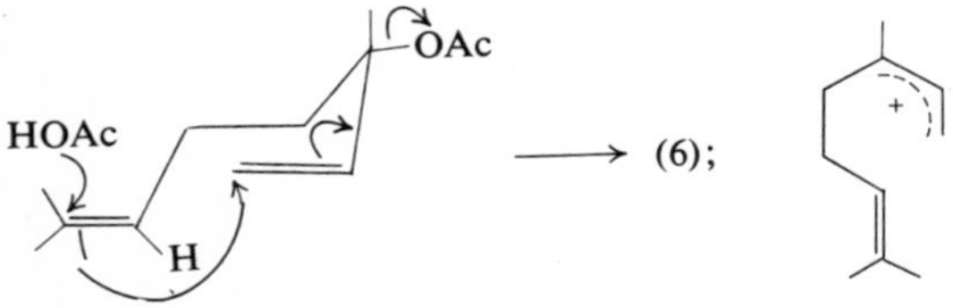

(7)

Since the product is found to be optically active, cyclisation of linalool (5) cannot involve an intermediate free carbonium ion (7) which would be racemic.

This general method of ring formation is discussed further in connexion with the synthesis of decalin derivatives (chap. 9).

1,5-Dicarbonyl derivatives may also be cyclised to give cyclohexanones. This route is illustrated in the synthesis of dimedone (8) *via*

Michael addition of sodio-malonic ester to mesityl oxide followed by Claisen condensation:

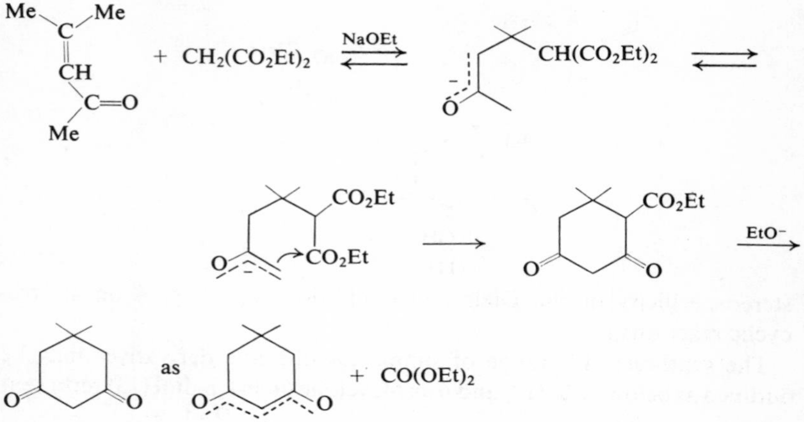

The Diels–Alder synthesis, which is a valuable route to six-membered rings, may be illustrated by the following examples:

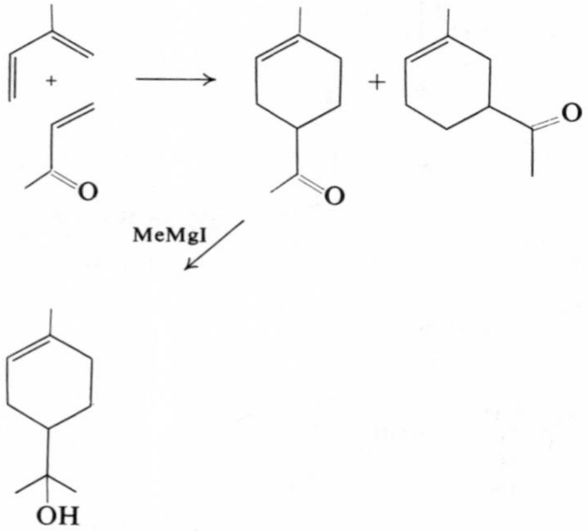

representing a synthesis of (±)-α-terpineol (9). Formation of the adduct (10) represents the valuable first step in a synthesis of shikimic acid (11) (Smissman *et al.*, 1959). The stereochemistry of (10) indicates the

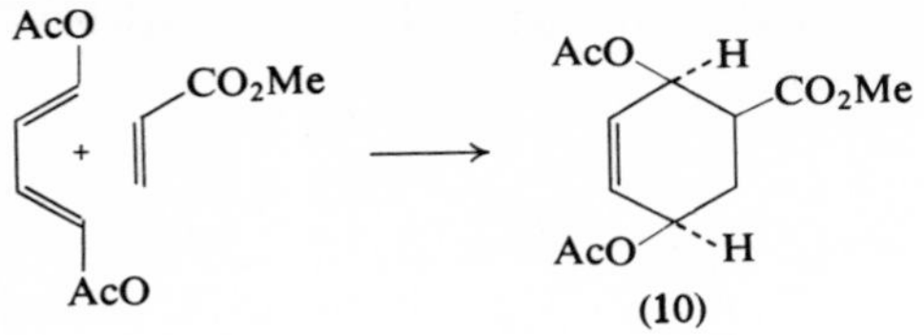

(10)

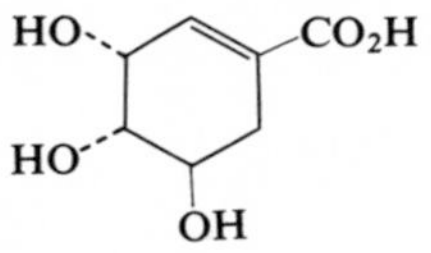

(11)

stereospecificity of the Diels–Alder addition (cf. chap. 4 on electro-cyclic reactions).

The synthesis of a range of simple cyclohexane derivatives may be outlined as below. *cis-* (12) and *trans-*Cyclohexane-1,2-diol (13) represent

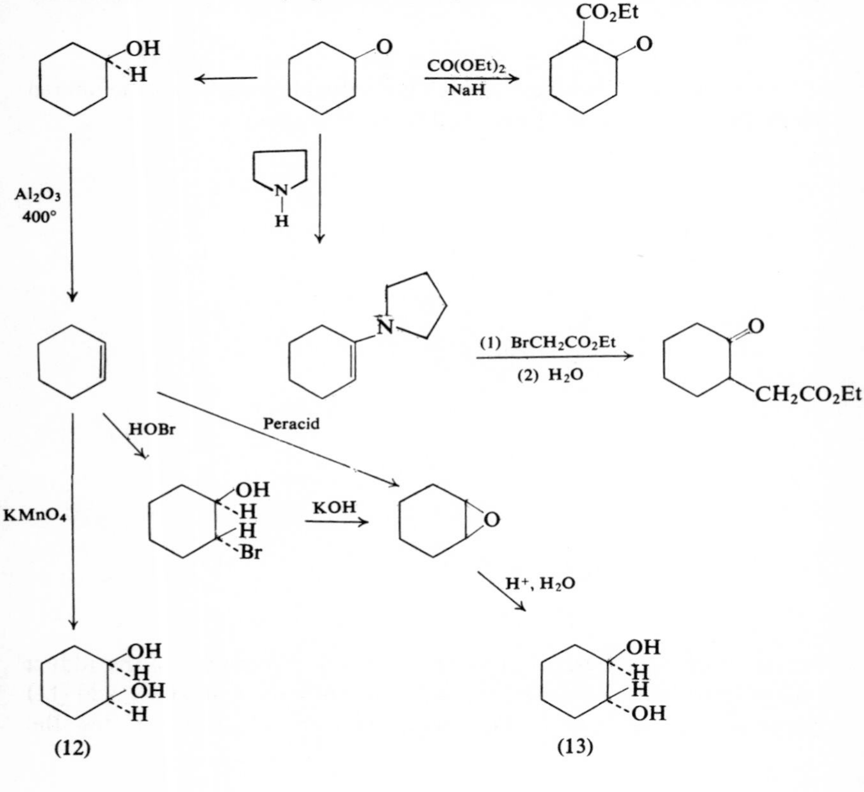

(12)   (13)

respectively *meso-* and (±)-forms, which are distinguished by resolution of (13) into (+)- and (−)-enantiomorphs.

6.3. Cyclohexenyl and cyclohexylidene derivatives. The study of cyclohexenyl (15) and cyclohexylidene (16) derivatives has provided interesting information regarding the relative stabilities of isomers, and the effects of strain originating through the interaction of groups in the allylic grouping (14), which is present in both (15) and (16).

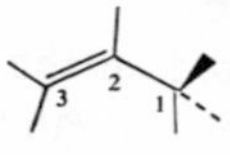

(14)

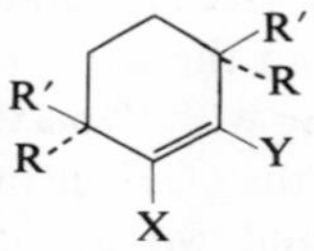

(15)

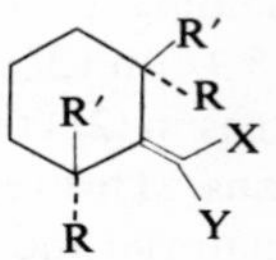

(16)

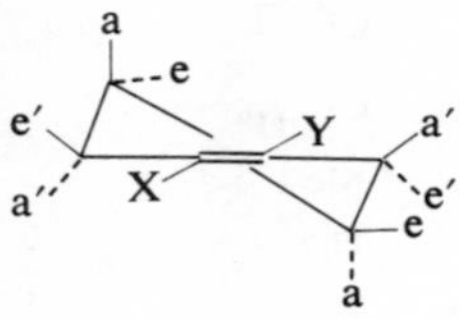

(17)

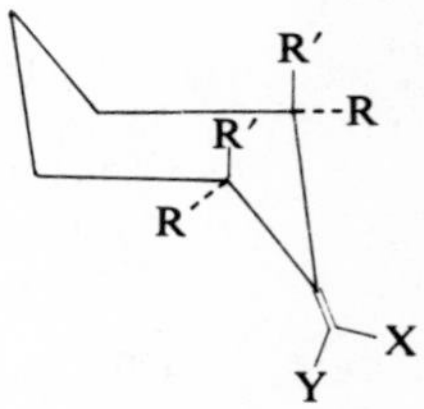

(19)

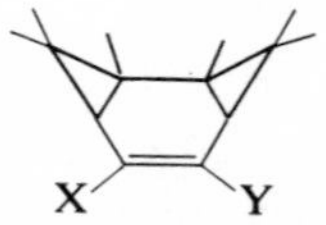

(18)

X-ray data show that the half chair (17) represents the preferred conformation for cyclohexene, although the energy increment in the half boat conformation (18), i.e. 11.3 kJ/mole, is only half that between the chair and the twisted boat conformations of cyclohexane (23.1 kJ/mole). A cyclohexene, (17), contains a planar group of four carbon atoms, ___/ and consequently the bonds marked e′ and a′ are bent away from an equatorial or axial orientation, and are designated quasi-equatorial and quasi-axial bonds respectively.

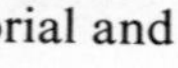

Inspection of (15) and (17) shows that the groups X or Y lie close to the adjacent e'-group R or R'. This interaction between 1,2-related groups introduces what has been designated as allylic-1,2- or $A^{1,2}$-strain, cf. (14). Similarly, (16) and (19) show interaction between X or Y and the 1,3-related equatorial R-group, which represents allylic-1,3-, or $A^{1,3}$-strain, cf. (14) (Johnson, 1968).

Clearly, an axial orientation for the larger of the groups R and R' may reduce this interaction, and hence will represent the preferred conformation in both (15) and (16).

$A^{1,3}$-strain may be illustrated in the case of (20) and (21). The *cis* isomer (20) is found to adopt the conformation (22) with the methyl group axially oriented since the methine hydrogen shows considerable deshielding by the proximate carbonyl group, viz.: $CH(CH_3)$ signal at τ 5.9. In (22) allylic coupling is seen between H^a and the axial H^b, whereas in (21) the signal for H^a indicates allylic coupling with two axial protons. This requires the conformation (23) in which the $-CH_3$ group is equatorial and allylic coupling is possible between H^a and two protons, H^b and H^c at a dihedral angle of $90°$.

(20)   (21)

(22)   (23)

The following data for the formation of isomeric enamines, enolates, enol ethers, and enol acetates, (25) and (26), from 2-methyl cyclohexanone (24) illustrates $A^{1,2}$-strain, and its dependence on the size of the interacting group X, cf. Stork and Hudrlik, (1968); House, (1965); Gurowitz and Joseph, (1967).

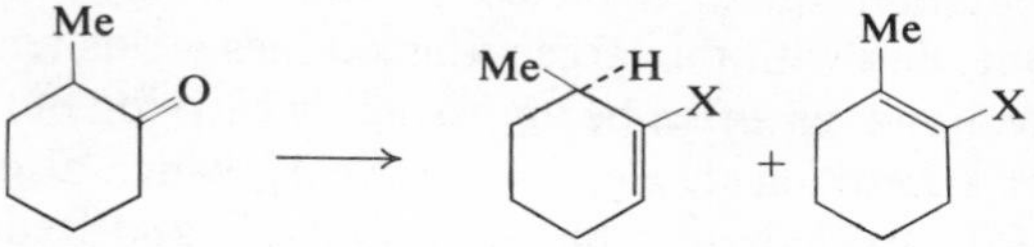

(24)   (25)   (26)

X	% (25)	% (26)
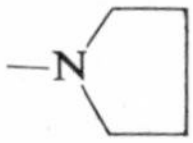 —N⟨	90	10
–OEt	51	49
–OLi	35	65
–OAc	5	95

Electronic factors which favour the more fully substituted olefinic isomer (26) are decisive with the enol acetate, but are counterbalanced by $A^{1,2}$-strain with increasing bulk of the X group, OAc < OLi < OEt < NR$_2$. The enolate group, OLi, will be solvated and the proportions of (25) and (26) are found to depend on the solvent. It is found also that in (25) the methyl group preferentially adopts the axial orientation.

6.4. Cycloheptane. Derivatives of cycloheptane are most frequently obtained by one of the methods of ring homologisation.

Cyclohexanone with diazomethane gives cycloheptanone in good yield, accompanied by some 10 per cent of an epoxide by-product:

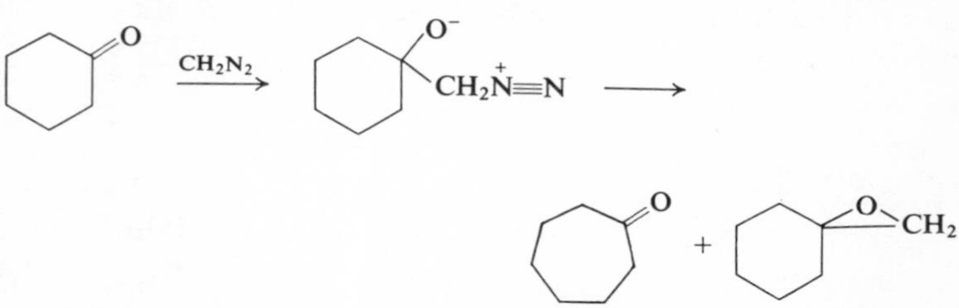

(27)

The use of this method depends, however, on the ketone product being less reactive towards diazomethane than the starting ketone. For this reason the preparation of cycloheptanone represents a favourable case; cyclopentanone gives a mixed product of cyclohexanone and cycloheptanone, and with the larger ring ketones yields tend to be low. The yield may, however, often be improved by carrying out the reaction in presence of a Lewis acid, e.g. BF_3 or $AlCl_3$, which also reduces the proportion of the oxide by-product (27) (Gutsche and Redmore, 1968).

The Demjanov–Tiffeneau ring expansion, based on deamination and pinacolic rearrangement is a useful general method (Gutsche and Redmore, 1968), giving good yields over a wide range of ring sizes, *viz*:

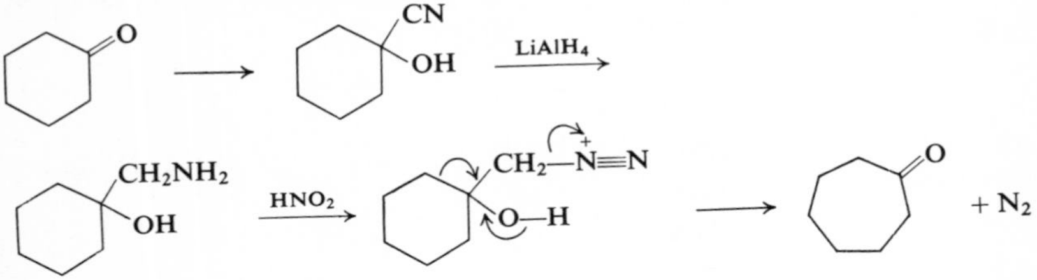

6.5. Cycloheptatriene. Cycloheptatrienes may be obtained by ring homologisation of an aromatic with a diazoalkane (Muller and Fricke, 1963), e.g.:

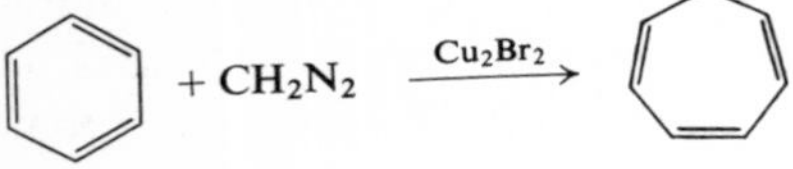

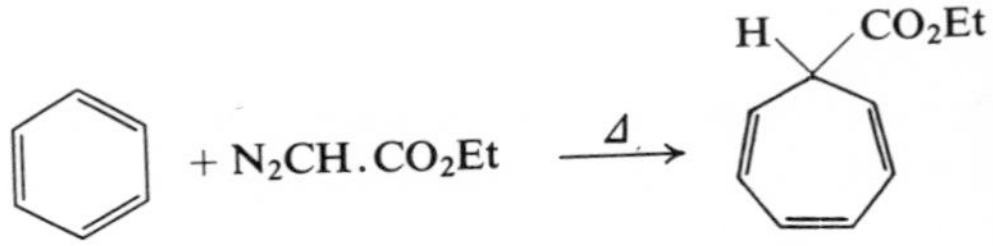

(and isomers)

or, as in the synthesis of the mould metabolite, stipitatic acid (28) (Bartels-Keith *et al.*, 1951):

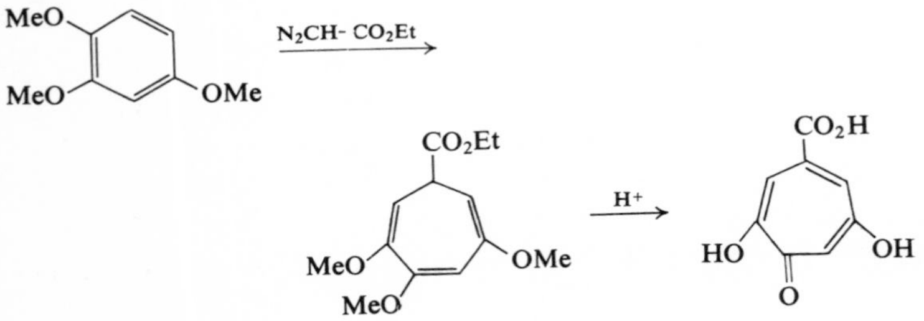

(28)

The following example (Nelson *et al.*, 1959), illustrates a convenient method based on ring expansion by Wagner–Meerwein rearrangement:

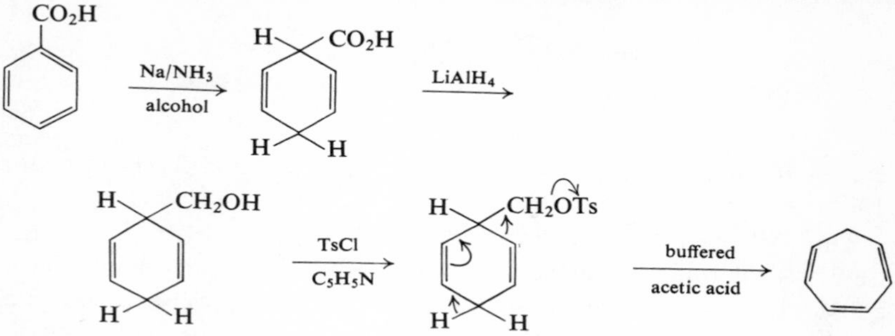

(TsCl = *p*-CH₃C₆H₄SO₂Cl

or, as applied to the synthesis of tropolone (29) (Chapman and Fitton, 1961; 1963).

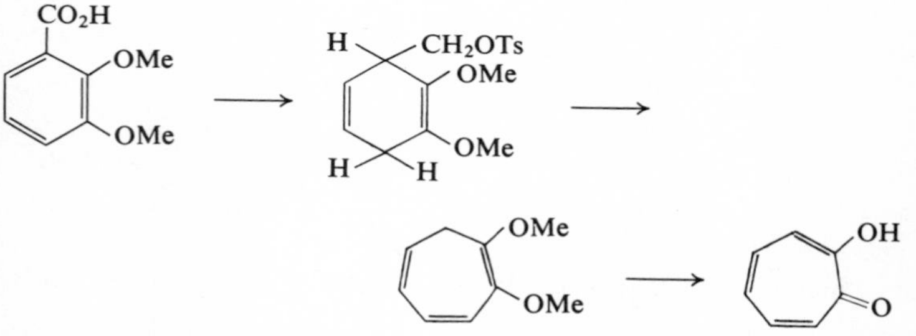

                                                                    (29)

The chemistry of cycloheptatriene (30) has a number of points of particular interest. The molecule is non-planar with rapid inversion between the equivalent structures (30) and (31). At low temperatures where

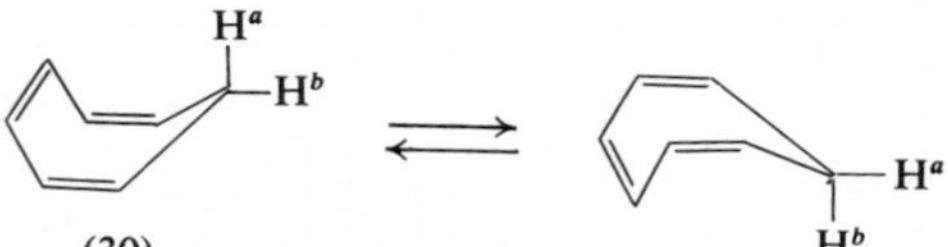

inversion is sufficiently slow the n.m.r. proton signals for H^a and H^b are resolved, and from the temperature dependence of the spectrum the activation energy for inversion: (30)⇌(31), is derived as 26.5 kJ/mole (Anet, 1964; Jensen and Smith, 1964).

Cycloheptatriene also exhibits a process of thermal isomerisation due to 1,5-hydrogen transfer across the molecule. 7-Deuterocycloheptatriene

(32), for example, is isomerised, and this process may be followed by the

(32)   (33)

change in the n.m.r. signal for allylic and vinyl protons (TerBorg *et al.*, 1963).

This hydrogen shift which is a concerted process of hydrogen migration may continue by further rearrangements of the same kind. As a consequence, the Buchner synthesis of cycloheptatriene carboxylic ester from benzene and ethyl diazoacetate leads to a mixture of isomeric

products which are interconverted by a sequence of 1,5-hydrogen shifts as indicated.

It is convenient to include here an analogous isomerisation of cyclopentadienes, by 1,5-hydride shift, which may be followed by the loss of

(34)   (35)   (36)

the τ 8.92 CH_3 n.m.r. signal in (34), and the development of a τ 8.05 CH_3 signal for the isomers (McLean and Haynes, 1964).

The rearrangements (32)→(33) and (34)→(35) and (36) are examples of what are designated by Woodward and Hoffmann (1970) as sigmatropic reactions, in which a sigma bond changes its position in an adjacent π-electron system. The transition state for such a reaction has been described in terms of the combination of the orbital of the migrating hydrogen atom with an extended allylic radical of orbital symmetry shown in (37). Clearly, to maintain positive overlap between the orbital of the migrating group and an orbital of the same phase of the radical is only possible between 1,5-related sites, on the same face of the molecule. Thus a cycloheptatriene or cyclopentadiene rearranges by 1,5-hydrogen shift across the face of the molecule, i.e. in the terminology of Wood ward and Hoffman, suprafacially.

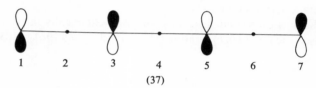

(37)

References

Anet, F. A. L. (1964). *J. Amer. Chem. Soc.*, **86**, 458.

Bartels-Keith, J. R., Johnson, A. W. and Taylor, W. J. (1951). *J. Chem. Soc.*, 2352.

Birch, A. J. (1950). *Quarterly Reviews*, 4, 69; (1958). *ibid.*, **12**, 7.

Chapman, O. L. and Fitton, P. (1961). *J. Amer. Chem. Soc.*, **83**, 1005; (1963). *ibid.*, **85**, 41.

Gurowitz, W. D. and Joseph, M. A. (1967). *J. Org. Chem.*, **32**, 3289.

Gutsche, C. D. and Redmore, D. (1968). *Carbocyclic Ring Expansions*, Academic Press.

House, H. O. (1965). *Modern Synthetic Reactions*, Benjamin, pp. 50, 64.

House, H. O. and Trost, B. H. (1965). *J. Org. Chem.*, **30**, 1341; 2502.

Jensen, F. R. and Smith, L. A. (1964). *J. Amer. Chem. Soc.*, **86**, 956.

Johnson, F. (1968). *Chemical Reviews*, **68**, 375.

Johnson, W. S., Bailey, D. Owyang, R., Bell, R., Jaques, B. and Crandell, J. (1964). *J. Amer. Chem. Soc.*, **86**, 1959.

Johnson, W. S. and Owyang, R. (1964). *J. Amer. Chem. Soc.*, **86**, 5593.

Komppa, G. (1903). *Chem. Berichte*, **36**, 4332; (1909). *Annalen*, **370**, 209.

McLean, S. and Haynes, P. (1964). *Tetrahedron Letters*, 2385.

Muller, E. and Fricke, H. (1963). *Annalen*, **661**, 38.

Nelson, N. A., Fassnacht, J. H. and Piper, J. A. (1959). *J. Amer. Chem. Soc.*, **81**, 5009.,

Smissman, E. E., Suh, J. J., Oxman, M. and Daniels, R. (1959). *J. Amer. Chem. Soc.*, **81**, 2909.

Stork, G. and Hudrlik, P. F. (1968). *J. Amer. Chem. Soc.*, **90**, 4462, 4464.

TerBorg, A. P., Kloosterziel, H. and Meurs, N. Van (1963). *Rec. Trav. Chim. Pays Bas*, **82**, 717; 741; 1189.

Woodward, R. B. and Hoffmann, R. (1970). *The Conservation of Orbital Symmetry*, Verlag Chemie/Academic Press.

7   Medium-sized and larger rings

7.1. Cyclo-octane derivatives, methods of synthesis. Dimerisation of buta-1,3-diene or tetramerisation of acetylene gives direct entry into the cyclo-octane series, cf. chap. 4.

Butadiene is dimerised thermally, or photochemically in presence of a sensitiser such as benzophenone (Hammond *et al.*, 1961, 1963, 1965)

the main product (1) from the thermal reaction representing the result of Diels–Alder addition whilst the major product from the photo-reaction is the divinyl cyclobutane (2) arising by a 2 + 2 cycloaddition process, cf. chap. 4.

cis-1,2-Divinyl cyclobutane (4) is easily rearranged thermally to cyclo-

octa-1,5-diene (5) and in the thermal dimerisation of butadiene some cyclo-octa-1,5-diene (5) is formed in this way (Vogel, 1963).

A more specific oligomerisation of butadiene is achieved *via* a Ni⁰-butadiene complex. This may be formed *in situ* by ligand displacement from, e.g. Ni(CH₂CHCN)₂, or from a Ni(II)-salt such as the acetate or acetylacetonate reduced *in situ* by means of an aluminium alkyl or a metal hydride. The reaction depends on formation of an intermediate of type (6) which may react with further butadiene to give *cis, trans, trans*-cyclododecatriene (7), or by addition of a phosphine undergo displacement to form cyclo-octa-1,5-diene, (5), cf. Wilke, (1963, 1970).

Suitable Ni(II)-complexes, e.g. the cyanide, acetylacetonate or salicaldimine derivatives also catalyse the polymerisation of acetylene to

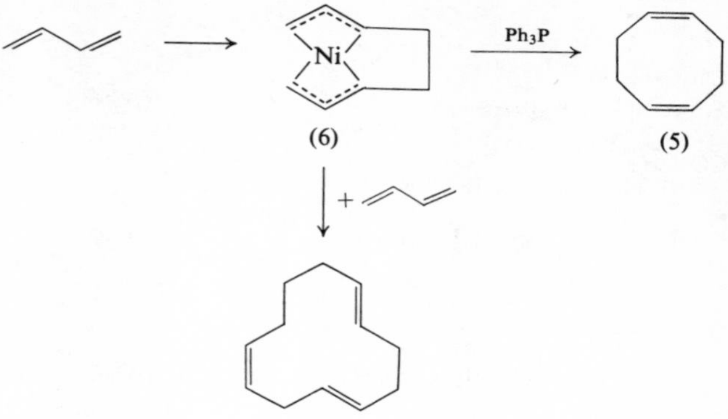

(6) (5)

(7)

give cyclo-octatetraene (9) in high yield (Schrauzer, 1964). This is represented as involving an intermediate nickel complex (8) bearing four co-ordinated acetylenes since by adding a phosphine, e.g. Ph₃P,

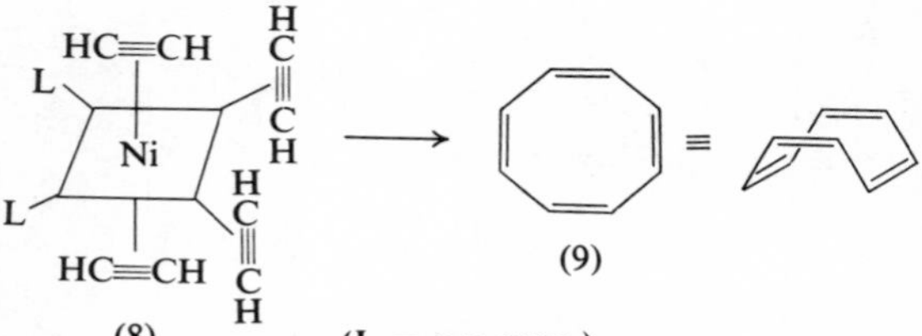

(8) (9) (L = e.g. acac.)

which blocks one co-ordination site, benzene becomes the main reaction product in place of cyclo-octatetraene.

Cyclo-octatetrene derivatives, e.g. (10) may also by synthesised by photoaddition of benzene and a suitable acetylene:

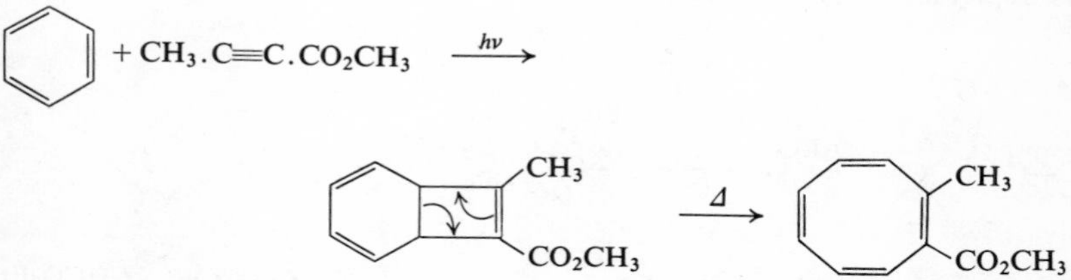

(10)

Another route is *via* photo-dimerisation of a cyclobutene-1,2-dicarboxylic ester followed by thermal rearrangement of the tricyclo-octane

(11)   (12)

intermediate (11) to the cyclo-octadiene tetracarboxylic ester (12).

Annelation followed by ring scission at a bridge is also of value in the synthesis of larger rings (Gutsche and Redmore, 1968*a*). The instance:

is dependent on β-keto-ester fission:

In a second example use is made of an enamine:

and ring scission is dependent on the fragmentation process:

7.2. Larger rings.   Homologation of cyclo-octene (13) by dibromo-

carbene addition, represents a useful route to a cyclononadiene (Gutsche and Redmore, 1968*b*) (14); isomerisation of the intermediate allene is to be noted (cf. chap. 1).

Ring expansion by two carbon atoms is possible by cycloaddition between a cycloalkenamine and an acetylenic ester *via* rearrangement of the intermediate cyclobutene:

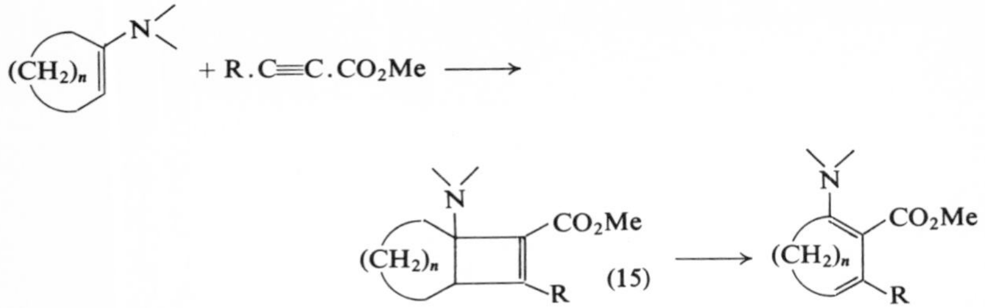

By a similar method a cycloheptenamine and methyl propiolate lead to a synthesis of cyclononanone (16):

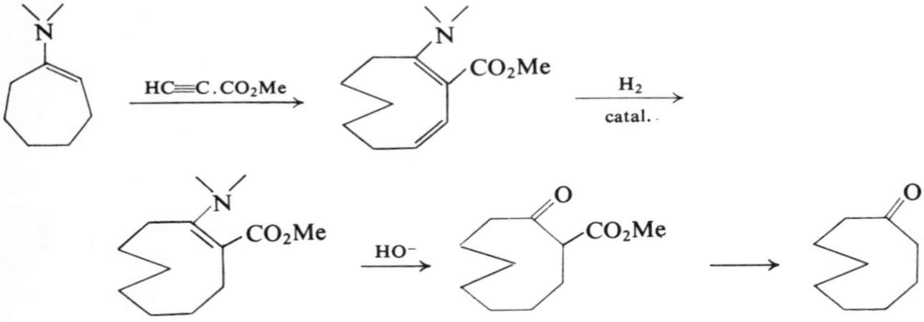

In many cases the cyclobutene type intermediate (15) is not observed, but is transformed directly to the ring expanded product (Gutsche and Redmore, 1968c). The cyclobutene → diene rearrangement is characteristic of cyclobutenes (cf. also chaps. 4 and 5).

7.3. The acyloin synthesis. A very widely used general method of synthesis of medium-sized rings is the acyloin condensation in which an α,ω-diester (17) is cyclised at the surface of metallic sodium to give an acyloin (18), as discussed in chap. 4 (McElvain, 1948; Sicher, 1962).

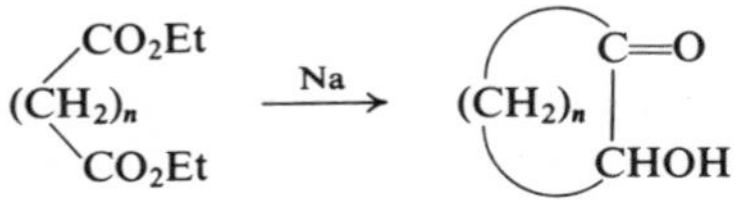

The acyloin (18) is also a convenient source of other derivatives (Blomquist *et al.*, 1952, 1953; Cope *et al.*, 1953, 1955; Prelog *et al.*, 1953).

The acyloin condensation gives yields of the order of 40–50 per cent in the synthesis of eight- or nine-membered rings, and with larger rings, e.g. for C-21-acyloin, yields may reach 90 per cent, cf. §4.2.

(a) CrO₃, (b) HgO/KOH treatment of the bishydrazone, (c) H₂/Pd, (d) Zn/HCl, (e) reductive amination of the ketone using CH₃NH₂ with hydrogen and nickel catalyst, (f) MeI, (g) Hofmann elimination; for proportions of *trans*- and *cis*-cyclo-alkene, cf. chap. 3.

7.4. Catenanes.   Use was made of the acyloin synthesis to prepare a product containing interlocking rings. Condensation of the ester (19) carried out by means of sodium under xylene in presence of the cyclic hydrocarbon (20) gave a product shown to contain the material (21), an example of what is called a catenane, since the molecule contains interlocking molecules as in the linking of a chain. The catenane (21) could be recognised by the presence of ν_{CD} bands in the infrared spectrum and by oxidation of the acyloin to release the deuterated hydrocarbon.

7.5. Ring inversion and Cope rearrangement of cyclopolyenes. Cyclo-octatetraene, cycloheptatriene and some related structures exhibit a range of rearrangement and other reactions of some general interest.

In cyclo-octatetraene the system of *cis*-olefinic bonds leads to a non-planar tub-like structure (22) with alternate single (146.2 pm) and double (133.4 pm) bonds. However, observation of the ^{13}C satellite n.m.r.

(22)   (23)   (24)

spectrum of cyclo-octatetraene has revealed a process of ring inversion in which (22) is converted into (24) *via* a planar structure (23) with an energy barrier of 57.5 kJ/mole (Anet *et al.*, 1962, 1964; Gwynn *et al.*, 1965).

In the isotopic molecule (25; $C^* = {}^{13}C$), H^a will be coupled with ^{13}C and with the adjacent H^b, whilst the dihedral angle of $c90°$ between H^a and H^c results in a little H^a–H^c coupling.

At $-53°$ the proton signal in the ^{13}C satellite spectrum appears as a doublet, but with rising temperature the signal broadens due to rapid (25)⇌(26) isomerisation (Anet *et al.*, 1964).

(25)   (26)

Cyclo-octatetraene also exhibits a process of isomerisation to the bi-cyclo[4,2,0]octatriene (28) which may be detected through the formation of Diels–Alder adducts such as (29). Kinetic studies indicate an equilibrium (27)⇌(28) containing at 100°C some 0.01 per cent of (28), (Huisgen, 1964) which has been independently prepared by low temperature dehalogenation of (30), but which rapidly forms (27).

Cycloheptatriene (31) behaves similarly in giving a Diels–Alder

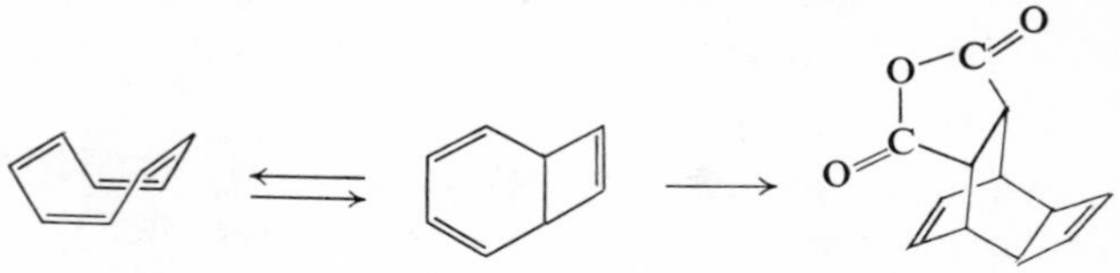

(27)   (28)   (29)

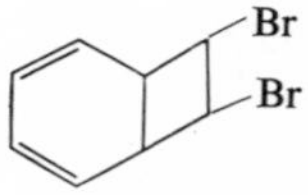

(30)

adduct (33) due to isomerisation to the bicyclo[4,1,0]heptadiene (32) which is trapped in the Diels–Alder reaction.

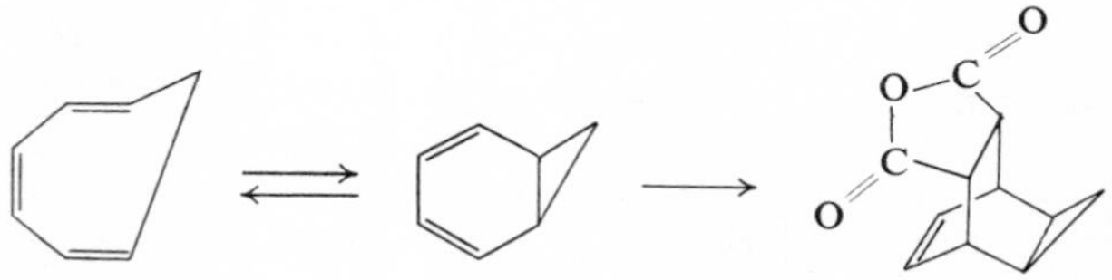

(31)   (32)   (33)

1,1-Dicyano-cycloheptatriene (34), on the other hand, exists prefer- entially as the bicycloheptadiene form (35), so that the equilibrium in this type of isomerisation is very sensitive to small structural change.

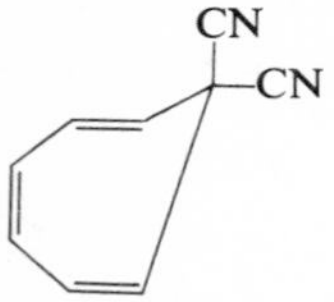

(34) 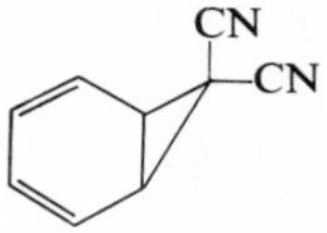   (35)

A similar instance is provided by cyclo-octa-1,3,5-triene (36) which equilibrates to give a relatively high proportion of the bicyclo-octadiene (37). Both (36) and (37) have been isolated (Cope *et al.*, 1952) and equilibrium re-established from either. (37) also forms Diels–Alder adducts across the cyclohexadiene system.

(36)   (37)

The isomerisation process: (27)→(28), (36)→(37), (31)→(32), and, e.g. the further example of *cis*-1,2-divinyl cyclobutane (38) (Vogel, 1963)

(38)   (39)

discussed earlier in this chapter, are examples of the Cope rearrangement (cf. Rhoads, 1964). This belongs to the class of electrocyclic reactions discussed in chap. 4, in which the termini of two π-electron systems interact by a concerted process: (38)→(39), to form a σ-bond, or conversely, in the reverse reaction a σ-bond breaks with formation of new π-bonds.

When the termini are not proximate, as in the case of *trans*-1,2-divinyl cyclobutane (40), this type of interaction is not possible, and (40)

(40)   (41)

is thermolysed to butadiene (41). This, however, requires a much higher temperature than is necessary for the concerted rearrangement of the *cis* isomer: (38)→(39) (Vogel, 1963).

In the case of the pyrolysis of the bis-amine oxide (42), the intermediate *cis*-divinyl cyclopropane (43) is spontaneously isomerised under the

(42)   (43)   (44)

conditions of the pyrolysis to give the cycloheptadiene (44) (Vogel, 1963).

7.6. Fluxional molecules. The initial and final structures of a Cope rearrangement may be identical, e.g. (*a*) and (*b*) in the case of bicyclo-[5,1,0]octadiene, (45), which is also known as homotropilidene:

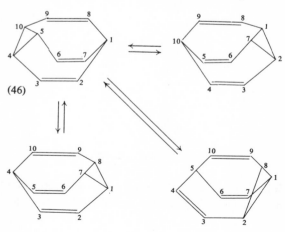

(a)   (b)

(45)

Homotropilidene (45) may be prepared by cuprous chloride catalysed addition of diazomethane and cycloheptatriene.

+ CH₂N₂ $\xrightarrow{\text{Cu}_2\text{Cl}_2}$ (45) +

At $-50°$ homotropilidene (45) shows a n.m.r. spectrum of cyclopropyl, allylic and vinylic protons in agreement with the structure (45). With rising temperature, however, the cyclopropyl signal is lost, the vinyl proton signal is much reduced, and the allylic protons are merged into a broad signal. At 180° the spectrum is again resolved with the development of a four-proton signal at τ 6.7 representing an average for protons between an allylic and vinylic situation. This change arises from very rapid rearrangement between (45) (a) and (b), and homotropilidene is therefore described as a fluxional molecule since at relatively low temperatures the structure is in a state of flux between equivalent structures (Doering and Roth, 1963; Schroder *et al.*, 1965).

Bullvalene (46), a derivative of homotropilidene but containing an extra olefinic bond, is also a fluxional molecule which, however, may be converted into equivalent structures in $\sim 10^6$ ways, e.g.:

(46)

Rapid interconversion between these valence-bond isomers results in a time-averaged structure in which all the protons become equivalent. At 100° bullvalene shows in the n.m.r. a single, strong signal at τ 5.8, but at $-25°$ vinylic (6H) at τ 4.3, and allylic (4H) protons at τ 7.9 may be recognised (Doering and Roth, 1963, *et al.*, 1967; Schroder *et al.*, 1965).

Bullvalene (46) is obtained from cyclo-octatetraene by thermal dimerisation:

followed by photolytic cleavage:

(46)

7.7. Homoaromatic structures.

In strongly acid solutions cyclo-octatetraene is protonated to the ion $C_8H_9^+$, which may be isolated as a salt $C_8H_9^+ SbCl_6^-$ from $HSbCl_6$ solution. The ion $C_8H_9^+$, named the homotropyllium ion (47), contains a group of atoms, 1–7, containing six π-electrons suitably placed for overlap in a planar delocalised manner as in the tropyllium ion (48). This group of six π-electrons represents an aromatic type of system on Huckel's definition of aromaticity.

(47)

(48)

(47) which is a representative of a group which Winstein (1969) has designated as homo-aromatics, shows aromatic type n.m.r. signals at τ 1.5

due to protons at 2, 3, 4, 5 and 6, and the inner proton H^a (τ 10.67) is strongly deshielded relative to the outer proton H^b (τ 4.9), in consequence of the aromatic ring current. The main signal at τ 1.5 is in the region to be expected for aromatic protons. The protons at 1 and 7 appear at τ 3.58.

The bishomocyclopentadienyl anion (49) and the trishomocyclopropenyl cation (51) are further examples of homoaromatic structures. The ion (49) is recognised from n.m.r. data, and there is evidence for (51) as a reaction intermediate in the acetolysis of the *cis*-bicyclo[3,1,0]-hexyl toluene-*p*-sulphonate (50).

References

Anet, F. A. L. (1962). *J. Amer. Chem. Soc.*, **84**, 671, cf. Anet, F. A. L., Bourn, A. J. R., and Liu, Y. S. (1964). *ibid.*, **86**, 3576.

Blomquist, A. T., Burge, R. E. and Sucsy, A. C. (1952). *J. Amer. Chem. Soc.*, **74**, 3636; Blomquist, A. T., Liu, L. H. and Bohrer, J. C., *ibid*, **74**, 3643; Blomquist, A. T. and Liu, L. H. (1953)., *ibid*, **75**, 2163.

Cope, A. C., Haven, A. C., Ramp, F. L. and Trembull, E. R. (1952). *J. Amer. Chem. Soc.*, **74**, 4867.

Cope, A. C., Pike, R. A. and Spencer, C. F. (1953). *J. Amer. Chem. Soc.*, **75**, 3212.

Cope, A. C., McLean, D. C. and Nelson, N. A. (1955). *J. Amer. Chem. Soc.*, **77**, 1628.

Doering, W. von E. and Potts, W. R. (1963). *Angewandte Chemie* (Internat. Edn.), **2**, 115; *Tetrahedron*, **19**, 715.

Doering, W. von E. *et al.* (1967). *Tetrahedron*, **23**, 3945.

Gutsche, C. D. and Redmore, D. (1968). *Carbocyclic Ring Expansions*, Academic Press, (*a*) p. 183, (*b*) p. 156, (*c*) p. 178.

Gwynn, D. E., Whitesides, G. M. and Roberts, J. D. (1965). *J. Amer. Chem. Soc.*, **87**, 2862.

Hammond, G. S., Turro, N. J. and Fischer, A. (1961). *J. Amer. Chem. Soc.*, **83**, 4674.

Hammond, G. S., Turro, N. J. and Liu, R. S. H. (1963). *J. Org. Chem.*, **28**, 3297, cf. Liu, R. S. H., Turro, N. J. and Hammond, G. S. (1965). *J. Amer. Chem. Soc.*, **87**, 3406.

Huisgen, R. (1964). *Angewandte Chemie* (Internat. Edn.), **3**, 83.

McElvain, S. M. (1948). *Organic Reactions*, **4**, 256.

Prelog, V., Schenker, K. and Kung, W. (1953). *Helv. Chim. Acta.*, **36**, 471.

Rhoads, S. J. (1964) in *Molecular Rearrangements*. Ed. de Mayo, P., Interscience,
p. 655.

Schrauzer, G. N. (1964). *Angewandte Chemie* (Internat. Edn.), **3**, 185.

Schroder, G., Oth, J. F. M. and Merenyia, R. (1965). *Angewandte Chemie* (Internat.
Edn.), **4**, 752, cf. *Chem. Berichte*, **97**, 3140; 3150.

Sicher, J. (1962) in *Progress in Stereochemistry*, Ed. de la Mare, P. B. D. and Klyne, W.,
Butterworths, **3**, 206.

Vogel, E. (1963). *Angewandte Chemie* (Internat. Edn.), **2**, 1.

Wilke, G. (1963). *Angewandte Chemie* (Internat. Edn.), **2**, 105; (1970). *Advances in
Organometallic Chemistry*, **8**, 29.

Winstein, S. (1969). *Quarterly Reviews*, **23**, 141.

8 Bridged rings and cage molecules

8.1. General. The study of bridged rings and cage molecules has provided results of great interest in relating ring strain and bond angles with chemical and physical properties. The following examples illustrate typical structures, and also exemplify the nomenclature (cf. §1.2) for bicyclic systems which specifies in descending order the number of carbon atoms in each chain intervening between the bridgehead positions.

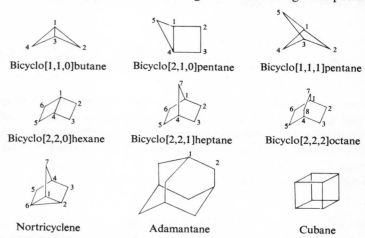

Bicyclo[1,1,0]butane	Bicyclo[2,1,0]pentane	Bicyclo[1,1,1]pentane
Bicyclo[2,2,0]hexane	Bicyclo[2,2,1]heptane	Bicyclo[2,2,2]octane
Nortricyclene	Adamantane	Cubane

8.2. Bicyclo[1,1,0]butane. This compound (1), and various functional derivatives (cf. Wiberg, 1968) have been obtained as follows:

133

Br—⬦—CO₂Me $\xrightarrow{\text{NaH}}$ ⬦—CO₂Me $\xrightarrow{\text{LiAlH}_4}$

⬦—CH₂OH

The chlorination by means of sulphuryl chloride is to be noted, and also the use of the Hunsdiecker reaction: $R.COOAg + Br_2 \rightarrow RBr + AgBr + CO_2$ for introduction of a bromo-substituent.

1,3-Disubstituted bicyclobutanes have been synthesised *via* the cycloaddition reaction:

$CH_2\!\!=\!\!C\!\!=\!\!CH_2 + CH_2\!\!=\!\!CHCN \longrightarrow CH_2\!\!=\!\!⬦\!\!-\!\!CN \xrightarrow{\text{HI}}$

$CH_3\!\!-\!\!⬦\!\!-\!\!CN$ (with I) $\xrightarrow[\text{ether}]{\text{NaH in}} CH_3\!\!-\!\!⬦\!\!-\!\!CN$

In the n.m.r. bicyclobutane shows two types of hydrogen substituent at τ 8.61 and τ 9.55 in the ratio 2:1. $J_{13_{CH}}$, the spin–spin coupling of a proton attached to ^{13}C isotopic carbon has been related to the s-character in the C–H bond (cf. chap. 1). The values of $J_{13_{CH}}$ in Table 8.1 provide an index of the state of hybridisation of different carbon hydrogen bonds in bicyclobutane in comparison with other similar rings.

TABLE 8.1   *Values of* J_{13C-H} (Hz) *in small rings*

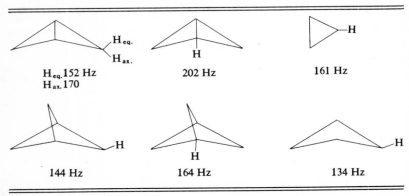

| $H_{eq.}$ 152 Hz $H_{ax.}$ 170 | 202 Hz | 161 Hz |
| 144 Hz | 164 Hz | 134 Hz |

In consequence of the differences in hybrid character of the bonds in bicyclobutane this molecule possesses a dipole moment; the microwave spectrum indicates a value of 0.67 D.

The bicyclobutane ring is very sensitive to acid hydrolysis, reacting 10^{10} times as fast as cyclopropane. From the enthalpy of the reaction

shown, the ring strain has been computed as 269 kJ/mole, i.e. more than twice the value for cyclopropane (116 kJ/mole) and 160 kJ/mole greater than for cyclobutane (109 kJ/mole). Acid fission of bicyclobutane gives both possible products, (2) and (3), in nearly equal amounts; in view of the ready interconversion of cyclobutyl and cyclopropyl methyl derivatives (cf. chap. 3) this is perhaps not unexpected.

The polarisability of bicyclobutane is evident in the ultraviolet absorption data:

The polarisability of bicyclobutane is evident in the ultraviolet absorption data:

slight absorption below 250 nm

λ_{max} 208 nm, ϵ, 12,500

Transmission of reactivity across the bicyclobutane ring is illustrated in the following examples which are reminiscent of β-addition to, e.g. acrylonitrile. Similar reactions have been noted with cyclopropanes carrying carbanion stabilising groups, cf. chap. 5.

Bicyclobutane is thermally decomposed to butadiene. The relatively large activation energy of 176 kJ/mole could be taken to indicate thermolysis *via* a radical intermediate:

but the process is found to be stereospecific (Closs and Pfeffer, 1968):

which suggests a concerted pathway:

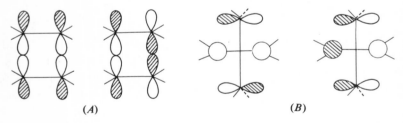

Concerted cycloaddition of two olefinic units in which bonding occurs by orbital overlap on the same face of the olefin molecule (suprafacially) as in (*A*) is symmetry disallowed (cf. chap. 4). However, Woodward has recognised a second mode of association of different geometry

(*A*) (*B*)

as in (*B*) in which the olefinic units approach orthogonally and bonding involves overlap between orbital lobes on opposite faces of one olefinic component, i.e. antarafacially. Process (*B*) is symmetry allowed, and thermolysis of bicyclobutane is regarded as the reversal of (*B*), in which both olefinic components are provided by butadiene. On this basis the decomposition of bicyclobutane is rationalised as a concerted stereospecific process (Woodward and Hoffmann, 1970).

8.3. Bicyclo[2,1,0]pentane. This compound (4) is obtained by a very generally useful route based on diene addition of cyclopentadiene and azodicarboxylic ester (Criegee and Rimmelin, 1957):

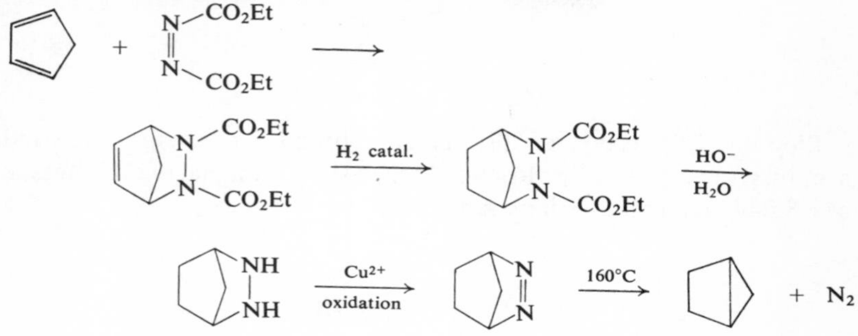

In place of azodicarboxylic ester, nitrosobenzene may be used as addend with cyclopentadiene. After hydrogenation the adduct is decomposed by heating (Griffin *et al.*, 1963):

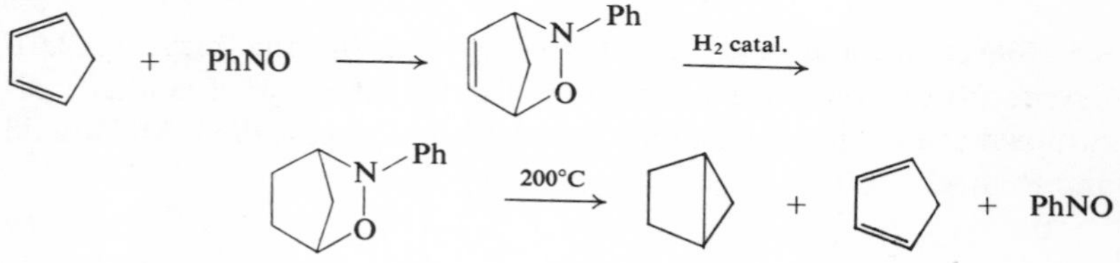

Bicyclopentane is thermally relatively stable; it is isomerised to cyclopentene at 350°. However, as a cyclopropane derivative, it reacts rapidly with hydrogen bromide to form cyclopentyl bromide.

8.4. Bicyclo[2,2,0]hexane. This product (5) may also be obtained *via* a nitrosobenzene adduct, in this case with cyclohexa-1,3-diene (Griffin *et al.*, 1963):

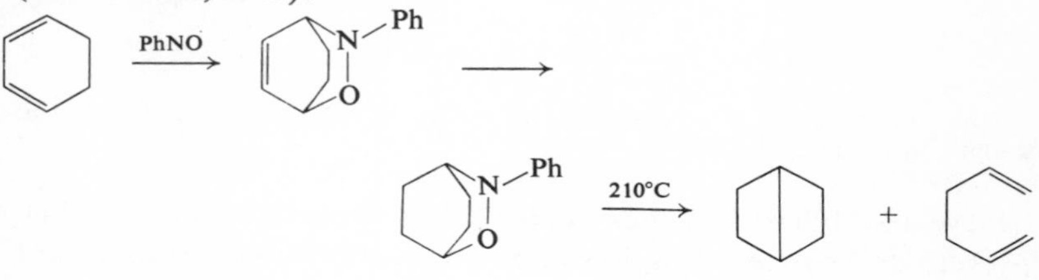

(5), 60%   (6), 30%

Bicyclo[2,2,0]hexane is, however, rather easily isomerised to hexa-1,5-diene (6), which is a by-product in the above synthesis:

(5)   (6)

Bicyclo[2,2,0]hexane, cf. (7), gives a broad n.m.r. signal centred around τ 7.66, i.e. at rather lower field than the CH_2 signal of cyclobutane at τ 8.04 (Cremer and Srinivasan, 1960).

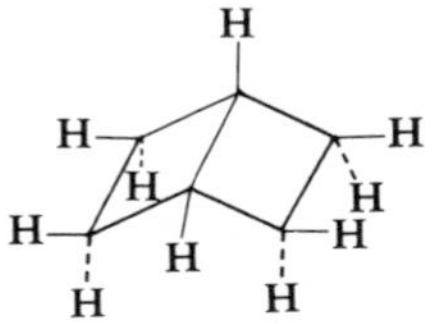

(7)

8.5. Bicyclo[2,2,0]hexadiene. This compound (8) and bicyclo[2,2,0]-hexene (9) are obtained by photocyclisation of a cyclohexadiene dicarboxylic acid anhydride (Van Tamelen and Pappas, 1963; McDonald and Reinecke, 1965):

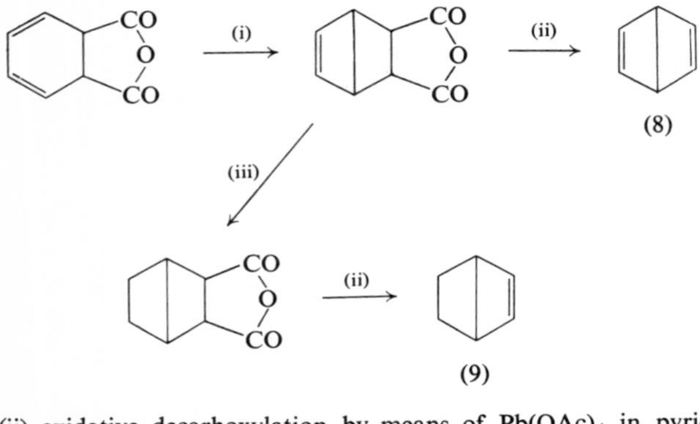

(8)

(9)

(i) $h\nu$, (ii) oxidative decarboxylation by means of Pb(OAc)$_4$ in pyridine, (iii) catalytic hydrogenation.

Bicyclo[2,2,0]hexadiene represents 'Dewar' benzene, and on keeping, it reverts to benzene. It is, however, remarkably stable having regard to the considerable ring strain. Retrogression to benzene requires appreciable activation energy. This arises because in terms of orbital symmetry correlation the process of isomerisation is symmetry forbidden (cf. chap. 4). The orbital symmetry correlations lead to benzene in an excited level, i.e. considerable activation is necessary.

Hexamethyl bicyclo[2,2,0]hexadiene, (10), is obtained, along with hexamethyl benzene, by an aluminium chloride catalysed trimerisation

(10)

of but-2-yne (Schäfer, 1966). Hexamethyl bicyclo[2,2,0]hexadiene has a strain energy of 187 kJ/mole, which, however, includes a small contribution from methyl group interactions around the ring. Hexamethyl bicyclohexadiene, on heating, reverts to hexamethyl benzene, but despite the release of ring strain this requires an energy of activation of 139 kJ/mole, i.e. this symmetry forbidden process is far from spontaneous (Oth, 1968).

8.6. Bicyclo[2,2,1]heptane. Norbornane (11) or bicyclo[2,2,1]heptane, is a structure of some interest based on a boat cyclohexane ring (atoms 1 to 6), and containing two cyclopentane units (1,2,3,4,7; and 4,5,6,1,7) bent along the 1,4-axis. Substituents at 2,3,5 and 6 may be *exo* or *endo* oriented, as shown.

$\overset{2}{>}C=O$ ν_{CO} 1751 cm^{-1}; $\overset{7}{>}C=O$ ν_{CO} 1773 cm^{-1}.

The degree of bond deformation in the structure (11), cf. Sim (1965) and Ching *et al.* (1968) is indicated by the accompanying bond angles based on X-ray data, and the values of ν_{CO}. The strain introduced by the 1,4-bridge may be estimated from the following heats of hydrogenation for bicyclo[2,2,1]hept-2-ene (12), bicyclo[2,2,2]oct-2-ene (13), cyclohexene, and cyclopentene. The bridging in (12), by compressing

−114 kJ/mole   −107 kJ/mole

(12)   (13)
−139 kJ/mole   −119 kJ/mole

the angles at 1:2:3 and 2:3:4 below the normal C—C=C angle of *c* 120°, raises the energy of the olefinic bond.

Similarly, in (11), the highly compressed 1:7:4 bond angle of 93° results in a strengthening of bonding of a group attached at 7, seen in the high carbonyl stretching frequency ν_{CO} for norbornan-7-one (14), i.e. 1773 cm^{-1}. The following rates of borohydride reduction ($k_2 \times 10^4$ l/mole s at 0°C) illustrate the same influence, i.e. reduction at C-7:

16.9   7.0

(14), 15 × 10^3 (15), 5.3

$>$C=O → $>$C$<^H_{OH}$ is assisted by strain release as the hybridisation changes: sp^2→sp^3. Norbornan-2-one (15), on the other hand, reacts at about the same rate as cyclopentanone (Brown and Muzzio, 1966).

In norbornane (11) the pairs of atoms: 2,3 and 5,6 form eclipsed ethane units in which compression on an *endo* is rather greater than on an *exo*-oriented group. *Exo/endo*-equilibrium data, indicate, however, that the difference is small, and indeed is negligible when one or more of the centres 2,3,5, or 6 is trigonally substituted (Cope *et al.*, 1959):

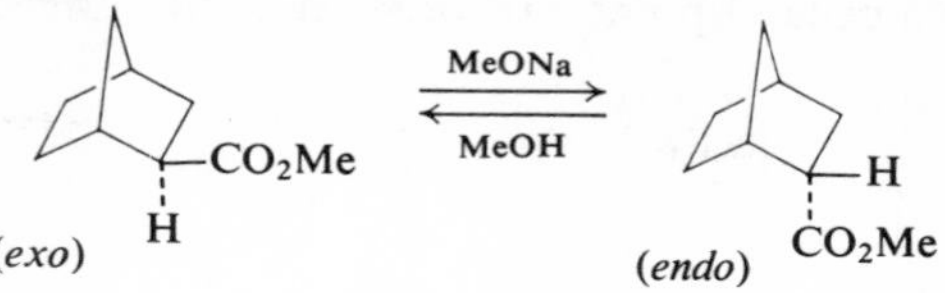

(*exo*) (*endo*)

K exo/*endo* = 2.4

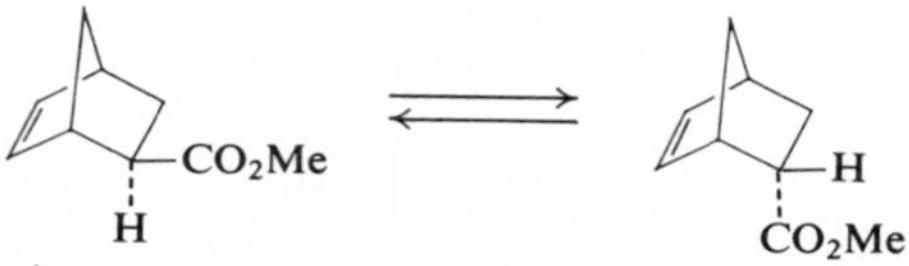

(*exo*) (*endo*)

K exo/*endo* ~ 1

In norbornane derivatives there is, however, a very marked preference for reaction from the *exo*-side. Thus reduction of norcamphor (15) with

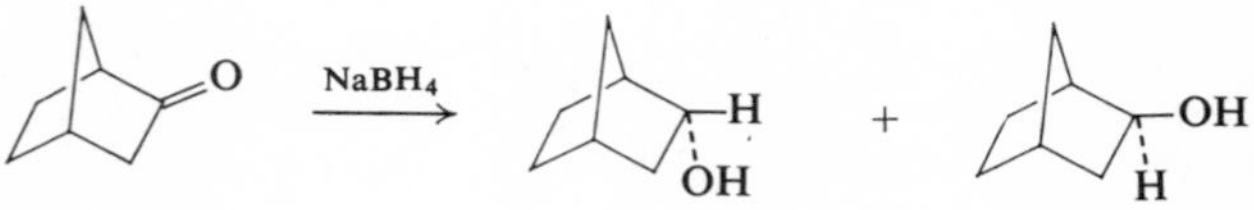

(15) (16), 86% (17), 14%

sodium borohydride gives mainly the *endo*-alcohol, norborneol (16), by *exo*-hydrogen addition and only a minor amount of the *exo*-alcohol (17) formed by *endo*-hydrogen addition. The same is true also of sodium/alcohol reduction: norcamphor (15)→77 per cent *endo*- (16), and 23 per cent *exo*-norborneol (17).

8.7. Synthesis of bicyclo[2,2,1]heptane and derivatives. Acetolysis of cyclopent-3-enyl ethyl toluene-*p*-sulphonate (18), which gives *exo*-norbornyl acetate (19) in good yield (Lawton, 1961), is a theoretically interesting synthesis of bicycloheptane. However, Diels–Alder addition

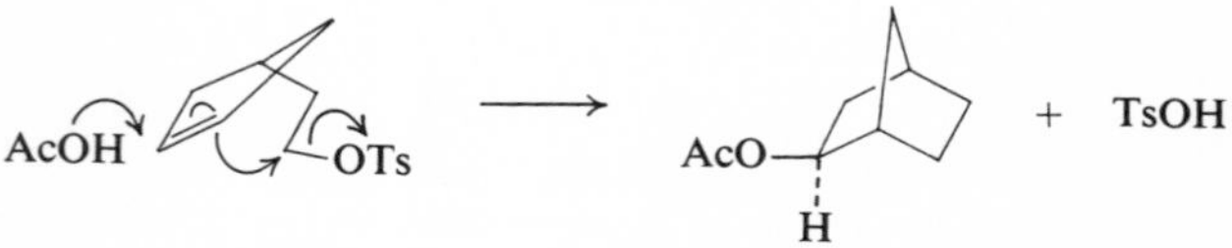

(18) (19)

to cyclopentadiene is a more generally useful method. The chart outlines a number of products derived in this way. These examples also draw attention to two general points: (i) predominantly *endo*-addition in the

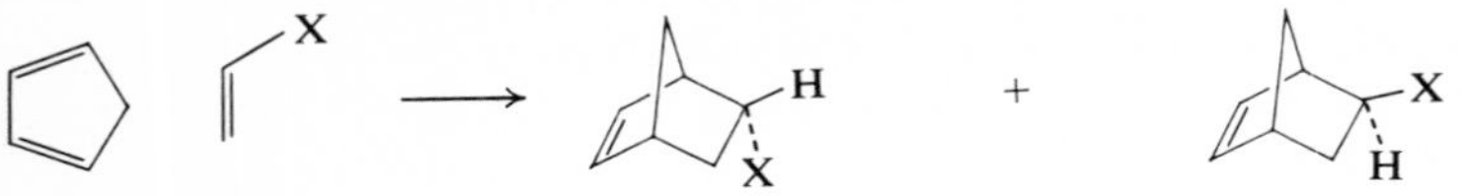

e.g., X = H, OAc, CO₂Me. For X = CO₂Me: *endo* (20):*exo* (21) = 76:24

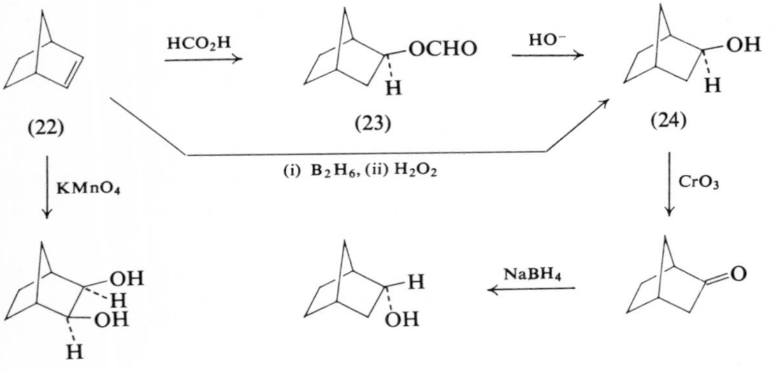

(22)   (23)   (24)

(i) B₂H₆, (ii) H₂O₂

(25)   (26)   (27)

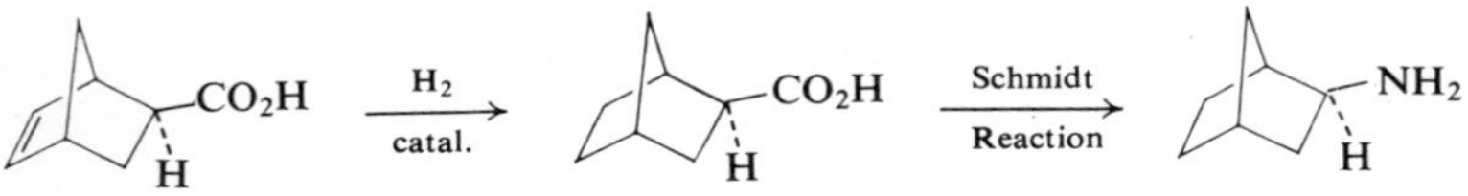

(28)

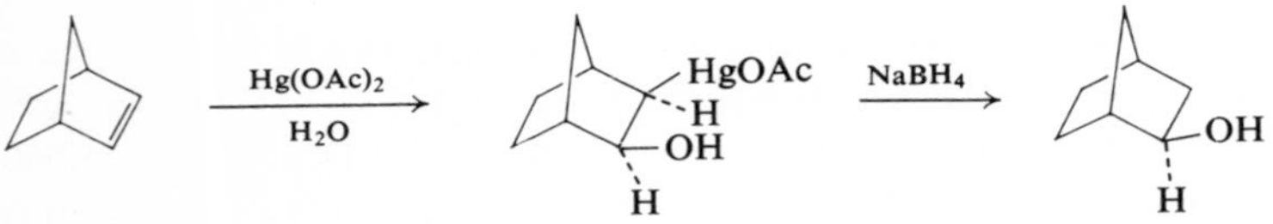

(22)   (29)   (24)

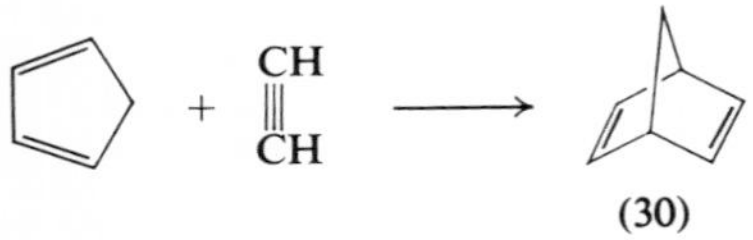

(30)

Diels–Alder reaction, and (ii) almost exclusively *exo*-addition to norbornene, e.g. (22)→(23), (22)→(24), (22)→(29) or to a norbornane
derivative: e.g. (27)→(26).

8.8. Bicyclo[2,2,1]hept-2-ene: addition reactions. The result of addition
of a reagent XY to norbornene depends on the nature of the reagent,
and is frequently accompanied by molecular rearrangement. Three
cases, (*a*), (*b*) and (*c*), outlined below, may be distinguished: (*a*) and (*b*)
where a relatively polar reaction complex leads to rearrangement, and
(*c*) where a less polar reaction complex results in formation of the XY-
exo-cis-adduct, i.e. without rearrangement. The results for different
reagents are summarised in Table 8.2 (Kaplan, *et al.*, 1960).

TABLE 8.2 *Products of addition of various
addends*, XY, *to bicyclo[2,2,1]hept-2-ene*

	Product		
Addend, XY	(31) %	(32) %	(33) %
HCO₃H	100	—	—
Cl₂ in pentane	40	60	—
Br₂/CCl₄/Pyridine	57	27	16
Hg(OAc)₂	—	—	sole product
NOCl	—	—	sole product
B₂H₆/H₂O₂	—	—	sole product
D₂/Catalyst	—	—	sole product

Case (c) represents an exception to the more familiar *trans-anti-coplanar* process of electrophilic addition which is general for most additions to cycloalkenes. However, in the process of *trans* addition to a cycloalkene as in (A), the angle θ is widened in forming the product. This angle widening is sterically possible in forming a cycloalkane derivative with a non-planar ring structure, e.g. a cyclohexane, or a cyclopentane or cyclobutane derivative where the ring may be twisted

(A)

somewhat. However, widening of the angle θ is resisted in a highly rigid structure such as a bicycloheptane, and bicycloheptene is considered to exhibit preferential *cis* addition, as in (B) for this reason (Traylor, 1968).

(B)

Similar considerations apply also to the reverse process of elimination, i.e. formation of a bicyclohept-2-ene should occur preferentially by loss of *exo-cis*-related groups. This is found to be the case. The dichloro-bicycloheptane (34) yields 2-chlorobicycloheptene much more readily than the isomer (35); (34) contains H– and Cl– groups *cis* related for elimination which is not the case in (35). Thus addition to an olefinic bond or elimination to form an olefin may occur preferentially by a

(34)   (35)

cis or *trans* process depending on the steric requirements of the ring system.

8.9. Nortricyclene. In the formation of the nortricyclene derivative (32) in (*b*) above, the angle compression which is associated with closing the cyclopropane ring is offset by release of angle strain in other parts of the structure. Thus bicycloheptene (36) and nortricyclene (37) have been interconverted (Schleyer, 1958) over a silica/alumina catalyst, and at

(36)   (37)

~100°C the equilibrium ratio: (36):(37) = 23:77 corresponds to a free energy difference of only 2.7 kJ/mole.

8.10. Norbornan-2-one and camphor. The following rate data ($k_2 \times 10^4$ l/mole s at 0°C) for sodium borohydride reduction indicate the difference in steric environment in norcamphor (27) and in camphor (38) due to alkyl substitution (Brown and Muzzio, 1966). The steric effect of the heavy alkyl substitution in camphor also results in preferred endo-

Rate 5.29, *exo/endo* = 6.2

(27)

Rate 0.0026, *exo/endo* = 0.16

(38)

reduction. By borohydride, and by catalytic reduction, camphor (38) gives mainly isoborneol (39) whereas from norcamphor (27) the *endo*-alcohol (26) is the main product of reduction:

(27)   (26)

(38)   (39)

In base catalysed proton/deuterium exchange, however, both camphor and norcamphor give exclusively the *exo*-monodeuteroketone
(Thomas *et al.*, 1967; Tidwell, 1970).

(38)

(27)

The difference between the stereochemistry of this reaction and of borohydride reduction is due to the difference in steric requirements of the
transition state of the two reactions. It has been suggested that in the
isotopic exchange specific *exo*-reaction of the enol (40) arises because the
torsional strain between the C—H^a and C—H^b bonds is reduced if the

(40)

addend group is *exo* and H^b moves into an *endo* orientation, but conversely is increased if the entering group approaches from the *endo* side
(Schleyer, 1967). The steric course of borohydride reduction, on the
other hand, appears to be dominated by steric hindrance in the parent
ketone.

8.11. The norbornyl cation. Displacement reactions of norbornyl
esters, particularly the toluene-*p*-sulphonates, have been the subject
of sustained research and controversy (cf. Brown, 1966). The main
facts may be summarised. (i) Both *exo*- (41) and *endo*-norbornyl-2-
toluene-*p*-sulphonate (42) on acetolysis give exclusively the *exo*-acetate
(43), the *exo*-ester (41) reacting *c.* 350 times as rapidly as the *endo*-isomer

(42). (ii) Acetolysis of optically active *exo*- and *endo*-norbornyl-2-*p*-bromobenzene sulphonates give an *exo*-norbornyl acetate totally racemic

(41)   (43)   (42)

from the *exo*-sulphonate, and at least 90 per cent racemic from the *endo*-sulphonate. (iii) Acetolysis of the ^{14}C-labelled ester (44, OBs = *p*-bromobenzene sulphonyl, $C^* = {}^{14}C$), gave an acetate with the ^{14}C-distribution shown (Roberts *et al*., 1954):

(44)

2,3 40% ^{14}C
1,4 23%
5,6 16%
7 21%

These results are best considered in terms of solvolysis *via* a norbornyl cation having a protonated nortricyclene structure such as (45*a*) or (45*b*). The energy of formation of the cyclopropane ring is minimised by strain release due to widening of angles in other parts of the structure and cyclopropanes are known to react by protonation. The ion (45*a* or *b*), which is symmetrical, may react at positions 1,2 or 6 to give the *exo*-acetate, which will necessarily be racemic and exhibit a redistribution of ^{14}C-label as is observed. The ions (45*a*) and (45*b*) represent respectively an edge-protonated and a corner-protonated cyclopropane, and (45*b*) is essentially the same as Winstein's non-classical norbornyl cation (46), proposed in 1949 as the intermediate in the solvolysis of norbornyl

(45*a*)

(45*b*)   (46)

sulphonate esters. The nature of this intermediate ion follows from spectroscopic study of the norbornyl cation which may be formed and examined at low temperatures in strongly acid solution.

An *exo*- or *endo*-norbornyl-2-halide (47a or b), or a norbornyl-1- or -7-halide, in SbF$_5$/liquid SO$_2$ solution at $-80°C$ give a common ion (48a or b) showing in the n.m.r. spectrum three groups of protons which are assigned as follows (Olah *et al.*, 1968, 1970; Schleyer *et al.*, 1964).

τ, TMS = 10:	8.14	7.18	4.99
Rel. intensity:	6	1	4
H at position:	3,5,7	4	1,2,6

That the n.m.r. data are not simply a time averaged spectrum from a set of rapidly interconverting classical carbonium ions, e.g.:

follows from the Raman spectrum (Olah *et al.*, 1968). This not only differs in general character from that shown by norbornane and norbornyl halides, but contains a ν_{CH} band at 3080 cm^{-1}, i.e. in the region characteristic of cyclopropyl CH. Moreover, nortricyclene in strong acid solution (FSO$_3$H/SbF$_5$/SO$_2$) shows the same type of spectrum. These findings are consistent with a protonated nortricyclene formulation such as (48a) or (48b) for the norbornyl cation intermediate in solvolysis. From recent study of ^{13}C n.m.r. spectra the corner protonated cyclopropane formulation (48b) emerges as the preferred structure (Olah *et al.*, 1970).

8.12. Homoenolisation. A tricyclene type of intermediate is encountered also in the racemisation of camphenilone (49) on treatment with potassium t-butoxide/t-butanol at $c.$ 250°. The reaction is represented as a process of homoenolisation by proton abstraction to give a symmetrical ion (50), which by reprotonation yields racemic camphenilone, (cf. Nickon and Lambert, 1966).

8.13. 7-Substituted norbornanes. 7-Substituted norbornanes are of interest because of the effect of the small 1:7:4 bond angle in reducing the reactivity of, e.g., a 7-toluene-p-sulphonate, and in increasing the reactivity of the 7-ketone. They are conveniently obtained by radical substitution of norbornadiene using t-butyl perbenzoate and Cu(I) bromide (Storey, 1961):

$$PhCO.OOCMe_3 + Cu^+ \longrightarrow PhCO_2Cu + Me_3CO.$$

(i) HOAc/HClO$_4$, (ii) Hydrogenation and hydrolysis, (iii) CrO$_3$/pyridine oxidation.

8.14. Nortricyclene and quadricyclene. Nortricyclene (51) may be prepared by oxidation of norcamphor hydrazone:

(51)

and norbornadiene (52) may also be cyclised by means of formic acid as follows

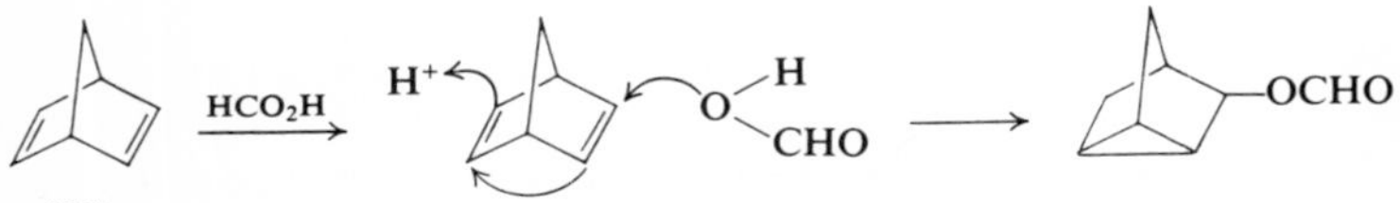

(52)

Photochemical cyclisation in presence of a sensitiser (e.g. Ph CO Me, or

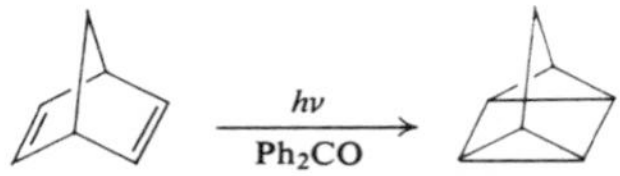

(52)   (53)

Ph₂ CO), gives quadricyclene (53) (Dauben, 1961). Reversion of quadricyclene (53) to norbornadiene (52), which is a symmetry disallowed process under the Woodward–Hoffman rules (cf. chap. 4), occurs only slowly on heating.

Norbornadiene (52) itself on strong heating (350–450°) forms cyclo-heptatriene (54) and toluene, as well as cyclopentadiene and acetylene (Herndon and Lowry, 1964):

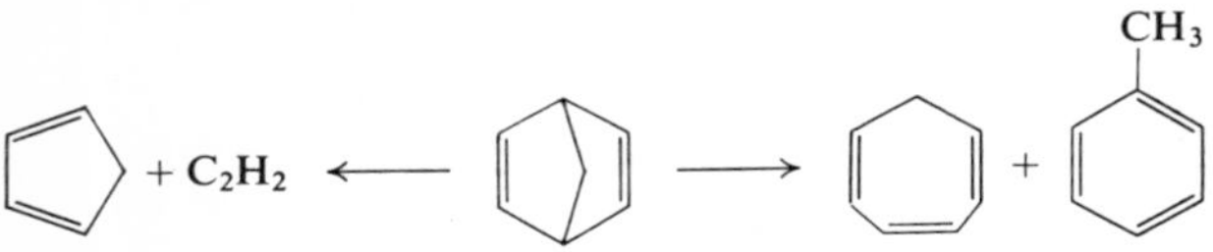

(52)   (54)

8.15. Bicyclo[2,2,2]octanes. These compounds (55) are obtained by Diels–Alder synthesis from cyclohexadiene (cf. Goering *et al.*, 1961).

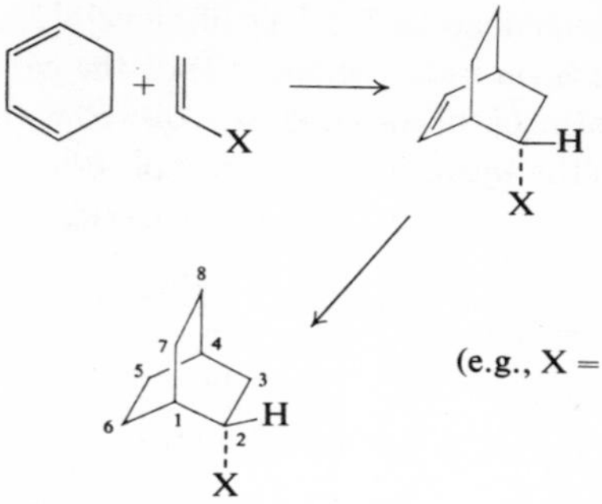

(e.g., X = H, OAc, CO₂Et)

(55)

Bicyclo-octane (55, X = H) is a symmetrical structure so that the distinction between *exo-* and *endo*-groups does not arise. As noted above, the heat of hydrogenation of bicyclo-octene is similar to the value for cyclopentene, and bicyclo-octan-2-one is of the same order of reactivity as cyclopentanone, i.e. bicyclo[2,2,2]octane is a relatively unstrained system. The toluene-*p*-sulphonate (56) is acetolysed at a rate intermediate between that for *exo-* and *endo*-norbornyl-2-toluene-*p*-sulphonates. The product, mainly bicyclo[2,2,2]octyl acetate (57) is accompanied by some 30 per cent of the bicyclo[3,2,1]octyl ester formed stereospecifically by Wagner–Meerwein rearrangement.

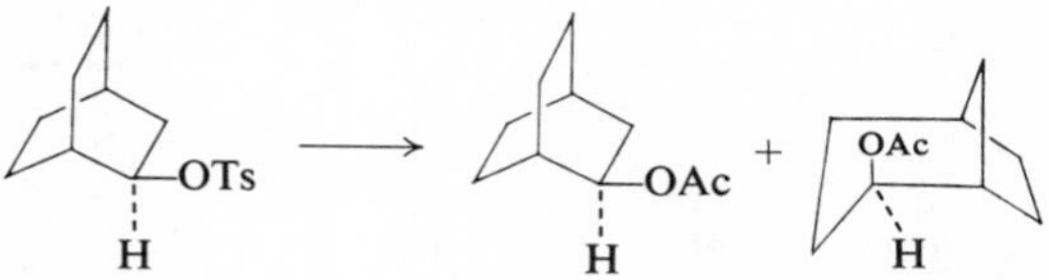

(56)   (57)

The reaction sequence:

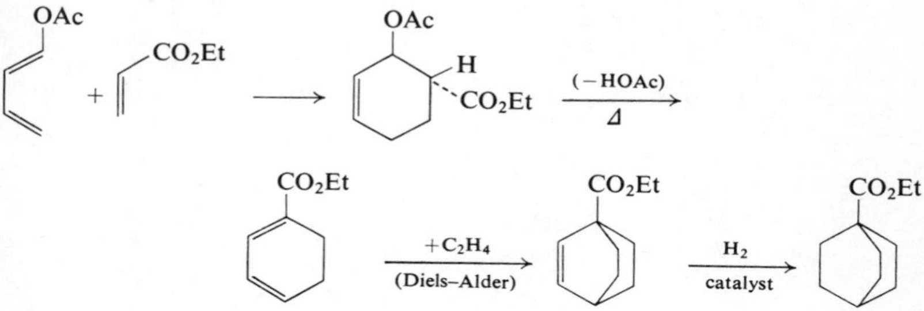

(58)

gives access to interesting bridgehead derivatives of bicyclo[2,2,2]octane,
e.g. (58*a*), and from this the bromo-derivative (59) or the carbinol (60).
The toluene-*p*-sulphonate of the carbinol (60) is solvolysed with Wagner–
Meerwein rearrangement to the bicyclo[3,2,2]nonanol (61).

8.16. Adamantane. This, which has the symmetrical and strainless
structure (63) is obtained from the dimer (62) of cyclopentadiene by

hydrogenation followed by the action of strong acids such as HBF_4, $HAlCl_4$, FSO_3H, or Lewis acids such as $AlCl_3$ (Schleyer, 1964). Adamantane undergoes bridgehead substitution, e.g. refluxing with bromine gives the 1-bromo-derivative (64) which by silver ion catalysed hydrolysis yields adamantan-1-ol (65) in high yield (cf. following section on bridgehead reactions).

The adamantyl-1-cation:

α	6H	τ 5.50
β	3H	τ 4.58
γ	6H	τ 7.33

is a structure of some stability which has been prepared in solution, e.g. from adamantyl fluoride and SbF_5, and characterised by its n.m.r. spectrum which is interesting in showing greater deshielding at the β than at the α carbon from the cationic centre (Olah, 1964).

8.17. 'Twistane'. This structure (66), isomeric with adamantane, has been synthesised by the interesting sequence (Whitelock, 1962):

Twistane is a dissymetric structure:

and, starting from the optically active carboxylic acid, (+)-twistane, $[\alpha]_D$ 414°, was obtained (Nakazaki, 1968).

8.18. Cubane. This structure (68) has been synthesised (Pettit *et al.*, 1966) in a very interesting manner by the decomposition of cyclobuta-diene iron tricarbonyl complex (67) in presence of 2,5-dibromo-1,4-benzoquinone. The released cyclobutadiene undergoes spontaneous addition to form an adduct, which, as in Diels–Alder addition, is the *endo*-isomer, for reasons of maximal orbital overlap. This adduct was further cyclised photochemically, and the derived dibromodiketone was converted into cubane dicarboxylic acid by Favorskii rearrangement.

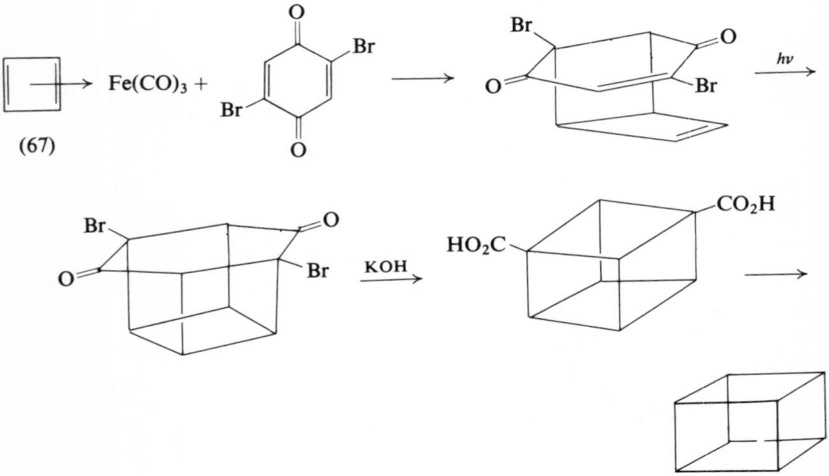

(67)

(68)

Decarboxylation was achieved by making use of the thermal decomposition of the t-butyl perester by heating in isopropyl benzene:

$$\geq C\!-\!COCl + HOOCMe_3 \longrightarrow \geq C\!-\!CO.OO.CMe_3 \longrightarrow$$

$$\geq C\cdot + CO_2 + Me_3.CO\cdot$$

$$\geq C\cdot + RH \longrightarrow \geq CH + R\cdot$$

$$(RH = solvent)$$

Cubane shows a single n.m.r. proton signal at τ 6, i.e. at very low field for a cycloalkane and $J_{13_{C-H}}$ of 160 Hz indicating the high s-character of the C–H bond due to angle strain in the ring.

8.19. Prismane. This compound (69) represents the prismatic structure once proposed for benzene by Ladenburg. Despite a strain energy of at least 380 kJ/mole, the prismane structure is remarkably stable (Oth, 1968; Woodward and Hoffmann, 1970).

Prismanes are known as various alkyl derivatives, obtained by photo-cyclisation of a bicyclo[2,2,0]hexadiene, e.g. hexamethyl prismane (70) from hexamethyl bicyclohexadiene (Schäfer *et al.*, 1967):

(69)   (70)

The hexamethyl bicyclohexadiene required is available from cyclo-trimerisation of but-2-yne, as was noted earlier in this chapter.

Hexamethyl prismane, which forms a crystalline solid, is isomerised on heating to reform hexamethyl bicyclohexadiene along with hexamethyl benzene in a ratio of *c.* 2.5:1, viz.:

However, whilst the photocyclisation used to form the prismane structure
is an allowed process in terms of the correlation of orbital symmetries, the
symmetry correlation for the thermal retrogression necessarily leads to
occupation of an anti-bonding level in the products. It is therefore easy
to understand the thermal stability of prismanes, and the high energy of
activation of *c.* 130 kJ/mole observed for the thermal decomposition of
hexamethyl prismane to hexamethyl-benzene and -bicyclo[2,2,0]-
hexadiene (Oth, 1968).

8.20. Bridgehead displacement reactions.   The relatively rigid bridged
structures of compounds such as adamantane or norbornane have
provided important information regarding the mechanisms of displacement of a group at a bridgehead position, and the extent to which such
displacement may be dependent on the ease of molecular deformation
at the reaction site.   The results may be considered in terms of the two
extreme cases: formation of a carbonium ion, and of a carbanion.

A carbonium ion is an sp^2 hybrid, and is ideally planar.   Hence in displacement at a bridgehead carbon, reaction in a strained structure may be

$$\theta\overbrace{}C-Y \longrightarrow C^+ \ Y^-$$

$$(71)$$

retarded both by compression of the angle θ, thus raising the s-character
of the C–Y bond (cf. chap. 1), and by structural resistance to reaching
planarity in the carbonium ion (71).   This effect is seen clearly in the
relative rates of solvolysis of the bromides:

$(CH_3)_3 CBr$

Rel. rate: 1 10^{-3} 10^{-6} 10^{-13}

These figures reflect the resistance to deformation of the bridged structures
to give a planar carbonium ion, and the extent to which deformation
towards planarity in the carbonium ion contributes to solvolysis. The
rates have been found to correlate with the calculated energy required
to deform the structures (Fort and Schleyer, 1967).

The strength of the C–Y bond is, however, also an important factor. Thus conversion of an alcohol into the chloride by treatment with thionyl chloride:

fails with norbornan-1-ol (72) since the intermediate chlorosulphite does not decompose:

(72)   (73)

to the halide (73), because of steric inhibition of the ionisation step:

There is, on the other hand, no difficulty about loss of nitrogen in the deamination of norbornyl-1-amine (74):

(74)

which gives a good yield of norbornyl-1-acetate. Similarly, the Hofmann, Curtius, or the Schmidt rearrangement:

$$R.CON_3 \longrightarrow R-N=C=O + N_2$$

and the Hunsdiecker reaction:

$$R.COOOAg + Br_2 \longrightarrow R.Br + CO_2 + AgBr$$

proceed normally where R = 1-adamantyl,1-bicyclo-octyl, or 1-bicycloheptyl.

In deamination, however, there is considerable driving force in the release of nitrogen which must contribute to forming the carbonium ion, whilst in the rearrangement reactions the migrating R-group remains effectively bonded during the migration.

It appears therefore that bridgehead displacement is seriously retarded in reactions where the energy of deformation from sp^3 towards sp^2 geometry at the reaction site makes an important contribution to the energy required for the displacement.

A converse situation arises when the reaction site retains four co-ordinate geometry, as, for example, in the formation of a metal alkyl derivative:

$$\theta \overset{|}{\underset{|}{C}}\text{—hal} \xrightarrow{\text{Li}} \overset{|}{\underset{|}{C}}\text{—Li}$$

In reactions of this type bridgehead displacement occurs readily, and the carbanion is formed more readily as the bridgehead bond angle θ is compressed, namely for reaction with lithium of the halides:

> (CH$_3$)$_3$C hal

i.e. as the s-character of the external bond is increased.

8.21. Bredt's rule and bridgehead double bonds. The existence and relative stability of substances such as cyclopropane, bicyclo[1,1,0]-butane, cubane or prismane illustrate the extent to which carbon–carbon σ-bonds may be bent yet maintain considerable orbital overlap and bonding as in (75). *trans*-Cyclo-octene (76) is a known, but highly strained and reactive olefin (cf. chap. 1) in which there must be consider-able twist about the olefinic bond as indicated by the arrows. From (77) it is evident that this twisting moment will reduce the π-bonding,

(75)   (76)   (77)

which is dependent on overlap of the p-lobes, and for which co-planarity is optimal.

These considerations are relevant to various reactions of certain bridged ring substances which are dependent on establishing an olefinic bond at a bridgehead position (Fawcett, 1950; Marshall and Fauble, 1970). Thus α-bromo-camphor (78), on treatment with base cannot be made to eliminate hydrogen bromide, i.e. H* is not removed with formation of an olefinic bond. Similarly bicyclo[2,2,2]octan-2,6-dione

(78)   (79)   (80)

(79) does not enolise as a 1,3-diketone, which would entail the development of double-bond character over the atoms 6,1,2. Further, ketopinic acid (80) fails to exhibit the easy decarboxylation, commonly typical of a β-keto-acid, which, however, also entails the introduction of an olefinic bond in forming the enolic intermediate in decarboxylation:

$+ CO_2$

These and similar observations led to the deduction by Bredt of the working rule that in small bridged ring systems it is not possible to introduce a double bond at the bridgehead carbon atom. However, it may be quite possible for the same ring system to accommodate without difficulty an olefinic bond in a non-bridgehead position. Thus the dehydro-camphoric acid (81) fails to form the anhydride (82) which contains a bridgehead double bond, but by isomerisation formation of the anhydride (83) is observed.

(82)   (81)   (83)

The projection (84) represents the relation of groups in ketopinic acid (80), viewed along the 2,1-bond. Decarboxylation:

could, in principle, release electrons into an orbital (*b*) as in (85). However, the orbital (*b*) is so related to the p-π-lobes (*a*) of the carbonyl group that overlap is not possible. Thus carbonyl assisted decarboxylation to the enol cannot occur and ketopinic acid does not decarboxylate. The same geometrical prevention of overlap is responsible for the non-elimination of hydrogen bromide from α-bromo-camphor (78) and the non-enolisation of the diketone (79). Thus Bredt's rule derives from

(84)   (85)

the rigidity of bridged ring structures which prevents the attainment of coplanarity for π-orbital overlap between adjacent carbon atoms necessary for generation of an olefinic bond at a bridgehead position.

It is relevant that whilst (86) represents the geometry of an olefinic group, (87), in which the substituents are in planes at right angles, is the geometry of the lowest excited state of an olefin. The energy, E, of twisting

(86)   (87)   (88)

an olefinic bond rises steeply with the angle of twist ω, cf. (88). For ethylene a relation: $E = 33.6 \ \omega^2$ J/mole has been used, which gives values of $E = 13.5$ and 30 kJ/mole for a twist of 20° and 30° respectively. Thus some degree of twisting is not energetically impossible, and Bredt's rule must be confined to small bridged rings where ω is necessarily large.

In agreement, the temperature for decarboxylation of the keto-acids (89), (90), and (91) falls with increasing ring size (Marshall and Fauble, 1970), and as the angle of twist about the derived enolic double bond falls, viz.: (89) 90°, (90) 66°, (91) 30°.

Also, it has been possible to synthesise the bicyclo[3,3,1]non-1-ene (92) as indicated, although this bridgehead olefin exhibits the reactivity towards addition reactions characteristic of a highly strained olefin.

The cyclisation of 1,5-diketones (93) (Prelog, 1950), discussed in chap. 9, also illustrates an aspect of Bredt's Rule since the proportions of the products (94) and (95) depend upon the ring size, and the yield of the isomer (95) carrying a bridgehead double bond in the bridged ring rises with increasing ring size, viz.:

n	% (94)	% (95)
4	65	0
5	32	14
6	0	76

References

Brown, H. C. (1966). *Chemistry in Britain*, 199.

Brown, H. C. and Muzzio, J. (1966). *J. Amer. Chem. Soc.*, **88**, 2811.

Ching, J. F., Wilcox, C. F. and Bauer, S. H. (1968). *J. Amer. Chem. Soc.*, **90**, 3149.

Closs, G. L. and Pfeffer, P. E. (1968). *J. Amer. Chem. Soc.*, **90**, 2452.

Cope, A. C., Ciganek, E. and Le Bel, N. A. (1959). *J. Amer. Chem. Soc.*, **81**, 2799.

Cremer, S. and Srinivasan, R. (1960). *Tetrahedron Letters*, No. 21, 24.

Criegee, R. and Rimmelin, A. (1957). *Chem. Ber.*, **90**, 414.

Dauben, W. G. and Cargill, R. A. (1961). *Tetrahedron*, **15**, 197.

Fawcett, F. J. (1950). *Chemical Reviews*, **47**, 218.

Fort, R. C. and Schleyer, P. von R. (1967). *Advances in Alicyclic Chemistry*, **1**, 284.

Goering, H. L., Greiner, R. W. and Sloan, M. F. (1961). *J. Amer. Chem. Soc.*, **83**, 1391; 1397.

Griffin, C. E., Hepfinger, N. F. and Sheperd, B. L. (1963). *J. Amer. Chem. Soc.*, **85**, 2683.

Herdon, W. C. and Lowry, L. L. (1964). *J. Amer. Chem. Soc.*, **86**, 1922.

Kaplan, L., Kwart, H. and Schleyer, P. von R. (1960). *J. Amer. Chem. Soc.*, **82**, 2341.

Lawton, R. G. (1961). *J. Amer. Chem. Soc.*, **83**, 2399.

Marshall, J. A. and Fauble, H. (1970). *J. Amer. Chem. Soc.*, **92**, 948, cf. Wiseman, J. R. and Fletcher, W. A. *ibid.*, 956.

McDonald, R. N. and Reinecke, C. E. (1965). *J. Amer. Chem. Soc.*, **89**, 3020.

Nakazaki, M. (1968). *Tetrahedron Letters*, 5467.

Nickon, A. and Lambert, J. L. (1966). *J. Amer. Chem. Soc.*, **88**, 1905.

Olah, G. A. (1964). *J. Amer. Chem. Soc.*, **86**, 4195.

Olah, G. A., Commeyras, A., and Liu, C. Y. (1968). *J. Amer. Chem. Soc.*, **90**, 3882.

Olah, G. A., White, H. M., DeMember, J. R., Commeyras, A. and Liu, C. Y. (1970). *J. Amer. Chem. Soc.*, **92**, 4627.

Oth, J. F. M. (1968). *Angewandte Chemie* (Internat. Edn.), **7**, 646; *Rec. Trav. Chim. Pays Bas*, **87**, 1185.

Pettit, R., Barborak, J. C. and Watts, L. (1966). *J. Amer. Chem. Soc.*, **88**, 1328.

Prelog, V. (1950). *J. Chem. Soc.*, 420.

Roberts, J. D., Lee, C. C., and Saunders, W. H. (1954). *J. Amer. Chem. Soc.*, **76**, 4501.

Schäfer, W. (1966). *Angewandte Chemie* (Internat. Edn.), **5**, 669.

Schäfer, W., Criegee, R., Askani, R. and Grunn, H. (1967). *Angewandte Chemie* (Internat. Edn.), **6**, 78.

Schleyer, P. von R. (1958). *J. Amer. Chem. Soc.*, **80**, 1700.

Schleyer, P. von R., Watts, W. E., Fort, R. C., Comisarow, M. B. and Olah, G. A. (1964). *J. Amer. Chem. Soc.*, **86**, 5679; 5680.

Schleyer, P. von R. (1964). *Chemical Reviews*, **64**, 277.

Schleyer, P. von R. (1967). *J. Amer. Chem. Soc.*, **89**, 701.

Sim, G. A. (1965). *J. Chem. Soc.*, 5974.

Storey, P. R. (1961). *J. Org. Chem.*, **26**, 287, cf. *Tetrahedron Letters*, 1962, 401.

Van Tamelen, E. E. and Pappas, S. P. (1963). *J. Amer. Chem. Soc.*, **85**, 3297.

Thomas, A. F., Schneider, R. B. and Meinwald, J. (1967). *J. Amer. Chem. Soc.*, **89**, 68, cf. Thomas, A. F. and Willhalm, B. (1965). *Tetrahedron Letters*, 1309.

Tidwell, T. T. (1970). *J. Amer. Chem. Soc.*, **92**, 1448.

Traylor, T. G. (1968). *Accounts of Chemical Research*, **2**, 152.

Whitelock, H. W. (1962). *J. Amer. Chem. Soc.*, **84**, 3412.

Wiberg, K. D. (1968). *Advances in Alicyclic Chemistry*, **2**, 185, cf. Wiberg, K. D., and Szeimies, G. (1970). *J. Amer. Chem. Soc.*, **92**, 573.

Woodward, R. B. and Hoffmann, R. (1970). *The Conservation of Orbital Symmetry*, Verlag Chemie/Academic Press.

9 Decalins, hydrindanes and related fused ring systems

9.1. Decalins. Decahydronaphthalene, or decalin, exists as *cis* (1) and *trans* (2) isomers. The principal steric relationships in the decalin series were established as follows (Hückel, 1925, 1926):

(i) Hydrogenation, e.g. Pt in acetic acid, or Raney Ni in alcohol; (ii) CrO_3; (iii) Clemmensen reduction; (iv) Oxidation; (v) Epimerisation by heating with hydrochloric acid.

The products belonging to the *trans* series were identified by the fact that the acid, (3), cyclohexane-*trans*-1,2-diacetic acid formed by oxidation, may be resolved into (+)- and (−)-forms; the corresponding *cis* acid (4) is the *meso*-isomer.

cis-Decal-1-one (5) was prepared as follows (Hückel, 1925).

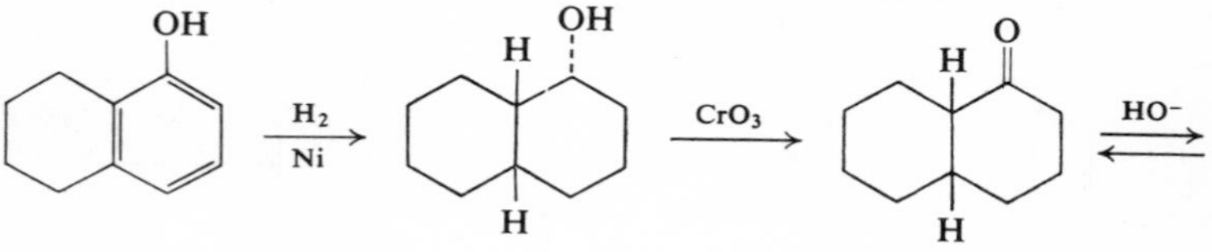

(5)

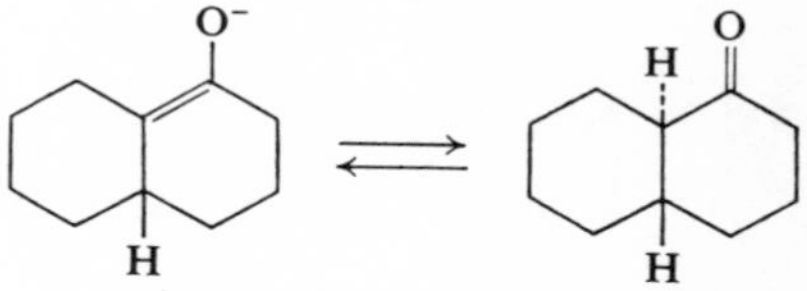

(5a)   (6)

It could be recognised as the *cis* form by alkali catalysed inversion to the more stable *trans* isomer (6) *via* the enolate (5a). Similarly, equilibration of the *trans*-decal-2-ols (7) and (8) (Hückel, 1963), gave a ratio (8):(7) =

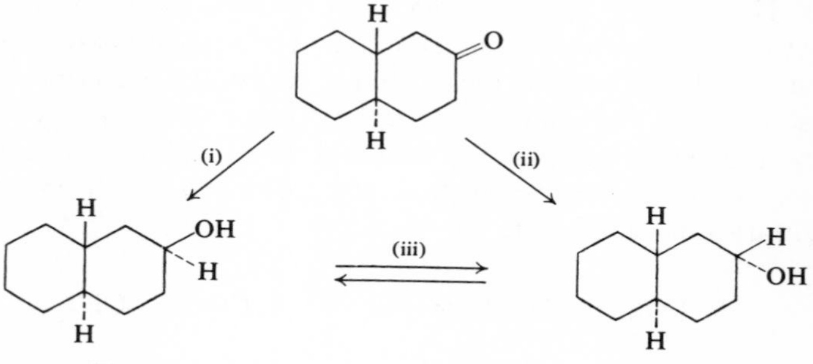

(7)   (8)

(i) H_2/Pt/HOAc; (ii) Na/alcohol, or $LiAlH_4$; (iii) Equilibration by means of Al(OPr^i)$_3$, or by heating with a nickel catalyst.

92:21, thus establishing (8) as the more stable (equatorial −OH) form.

9.2. Methods of synthesis. The above preparations of the decalols and decalones illustrate the use of an aromatic starting material. However, more general methods of synthesis of decalin derivatives have been developed and explored, particularly in relation to synthesis in the steroid and terpene group.

The following example, which represents the first stage in Woodward's cholesterol synthesis, illustrates the use of Diels–Alder addition. The

(9)

(10)   (11)

adduct (9) could be isomerised by enolisation with alkali to obtain the *trans* isomer (10) which was transformed into (11) (Woodward *et al.*, 1952). However, (9) may be noted as an instance where, owing to the relative planarity of the structure, the *cis* and *trans* isomers are of similar stability; equilibration gave only *c.* 40 per cent of the *trans* isomer (10).

The Robinson annelation method (cf. House, 1965) has been very widely used in the synthesis of octalones such as (13) and more complex derivatives. A vinyl ketone which may be generated *in situ* from a β-keto quaternary ammonium salt:

$$R.COCH_2\,CH_2N^+Me_3\,hal^- \xrightarrow{\text{Base}} R.COCH{=}CH_2$$

undergoes Michael addition with an enolate to give a 1,5-diketone which is cyclised to the ketol (12) by the alkoxide base present:

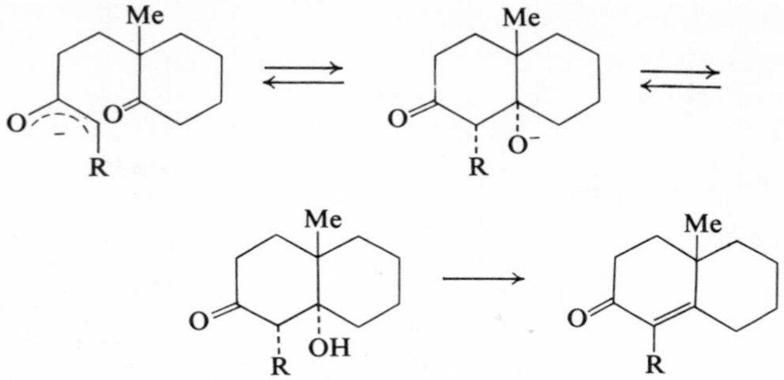

(12)   (13)

An alternative mode of cyclisation of the 1,5-diketone may arise in some instances (Johnson *et al.*, 1960):

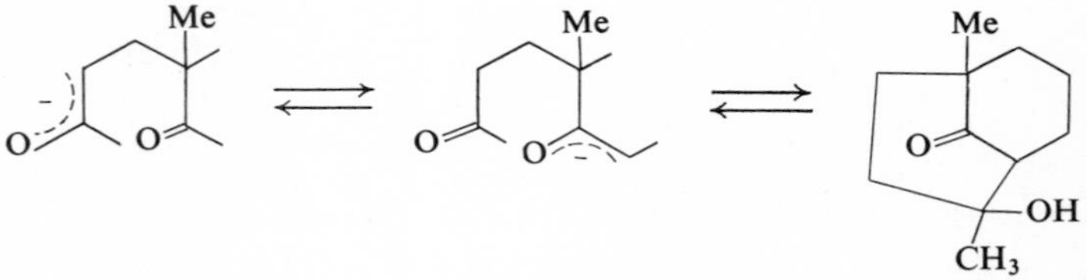

(14)   (15)

but is not normally important.   Inspection of the conformations corresponding to (12) and (15), viz.:

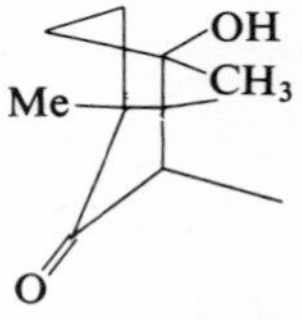

indicate why cyclisation as in (12) represents a preferred route; (15) involves more gauche bond interactions. However, when the Robinson annelation is applied to larger rings this steric situation is reversed (Prelog, 1950) and in the following example, (16), becomes the main product for values of $n = 5$ or larger (cf. chap. 8 concerning Bredt's rule), the CO_2Me group being lost during the reaction sequence:

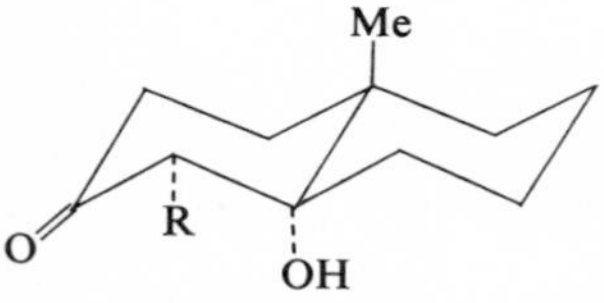

$$\begin{array}{c} CO_2Me \\ | \\ CH \\ \diagup \quad \diagup \\ (CH_2)_n \quad CO \\ \diagdown \quad \diagup \\ CH_2 \end{array} \quad + \quad \begin{array}{c} CH_2{=}CH \\ CH_3 \diagdown \diagup \\ O \end{array} \quad \longrightarrow$$

$$\begin{array}{c} CH{-}CH_2 \\ \diagup \quad \diagdown \\ (CH_2)_n \quad CO \quad CH_2 \\ \diagdown \quad \diagup \quad \diagup \\ CH_2 \quad CO \\ | \\ CH_3 \end{array} \quad \longrightarrow \quad \begin{array}{c} CH{-}CH_2 \\ \diagup \quad \diagdown \\ (CH_2)_n \quad CO \quad CH_2 \\ \diagdown \quad \diagup \quad \diagup \\ C{=}C \\ \diagdown CH_3 \end{array}$$

(16)

The decalin ring system may also be generated by acid catalysed cyclisation of a suitable 1,5,9-triene (Johnson, 1968), e.g.:

$$\xrightarrow[\text{HOAc}]{\text{H}_2\text{SO}_4}$$

However, in this method a variety of products, including double-bond isomers, and monocyclic products are normally formed, and yields are frequently low. A variation in which an epoxy-group replaces the terminal olefins (Van Tamelen *et al.*, 1963), viz.:

$$\xrightarrow{\text{BF}_3 \cdot \text{Et}_2\text{O}}$$

constitutes an interesting chemical analogue of the biochemical cyclisation of squalene oxide to lanosterol.

This type of cyclisation based on a sequence of ring closures has been considerably developed notably by W. S. Johnson, making use, for example, of an acetal to generate the electrophilic centre to initiate reaction. This type of cyclisation is found to be highly spectrospecific; in the example successive *trans-anti-trans* reaction at the olefinic bonds leads to the isomer shown. Further, by use of an acetal derived from an optically active glycol, e.g. butane-1,2-diol the new ring system is generated as a single enantiomer, e.g. (Johnson *et al.*, 1968):

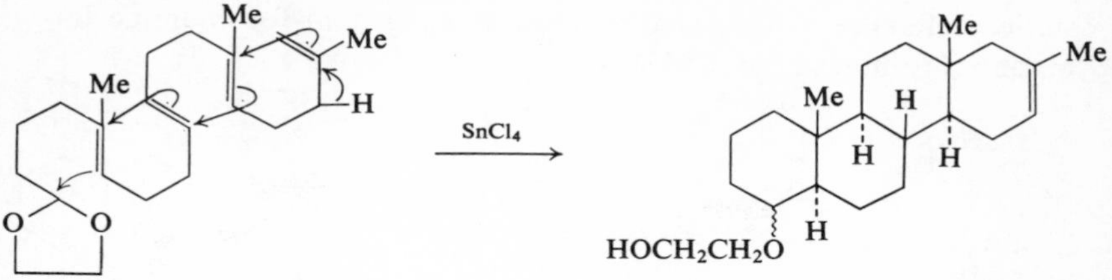

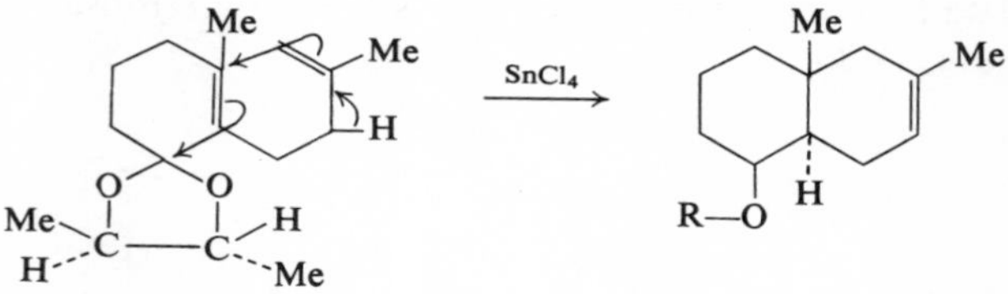

9.3. cis- and trans-Decalones. Reduction of the octalone available from the Robinson annelation may give access to either the *trans-* or *cis*-decalone series, e.g. in a simple instance:

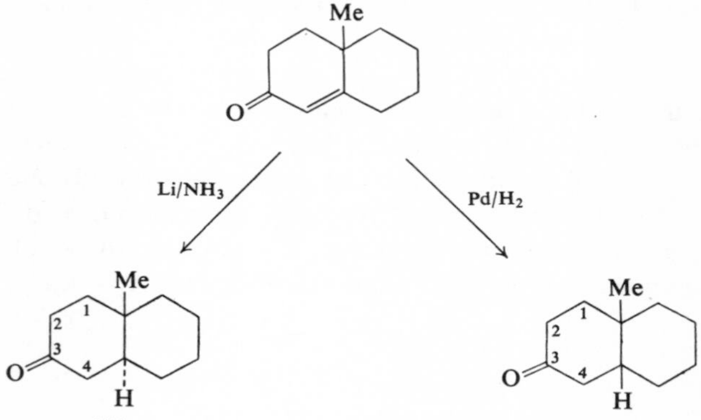

An interesting difference arises in the substitution reactions of *cis-* and *trans*-decal-3-ones. It is found that in reactions dependent on enolisation, e.g. bromination, or in enamine formation, molecules containing the *trans-* (17, 17*a*), or *cis-* (18) -decalone system react preferentially with proton loss from positions 2 or 4 respectively. This steric preference is indicated in the following data for enamine formation (Malhotra *et al.*, 1967):

(17*a*)   (19), 72%   (20), 28%

Similar data for enolisation illustrate the same point (House and Trost, 1963; House, 1965):

(18*a*)   (21), 32%   (22), 68%

(19) is stable relative to (20) since (20) contains a severe 4H:6αH interaction† and in (19) 4:6-interaction may be reduced by twisting of the ring structure. In the *cis*-decalone (18*a*) enolisation as in (22) removes 4α-:7α- and 4α-:9α-interactions which are present in (18*a*) and also in (21).

9.4. Decalones with a boat-chair conformation.

In chap. 2 reference was made to decalin derivatives in which steric interactions make a boat-chair type of conformation the preferred form (Robinson and Theobald, 1967). In the case of the tetrahydrosantonic acid (23), for example, a boat-chair conformation (25) obviates the 4αMe:6αOH interaction present in the chair-chair arrangement (24).

(23)

† The indication of steric orientation of groups by the prefix α or β is discussed in §1.2.

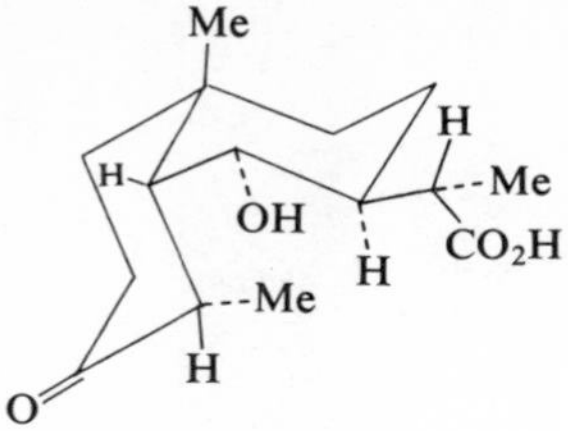

(24)

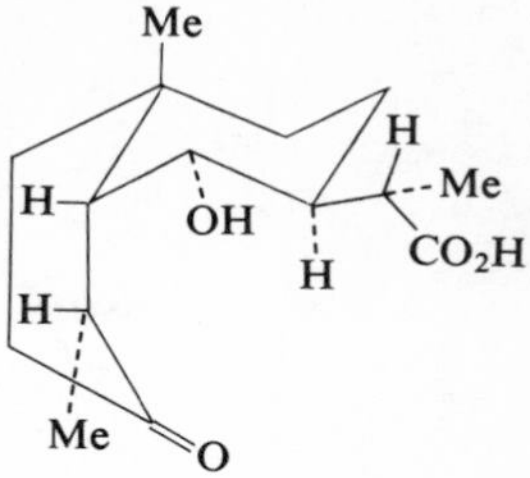

(25)

Similarly, (28) represents the preferred conformation of *trans*-tetra-hydroeremorphilone (26), since the chair-chair alternative (27) contains a severe 2β i-Pr:10βMe interaction. In (23) and (26) the presence of

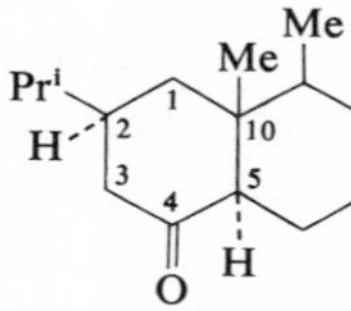

(26)

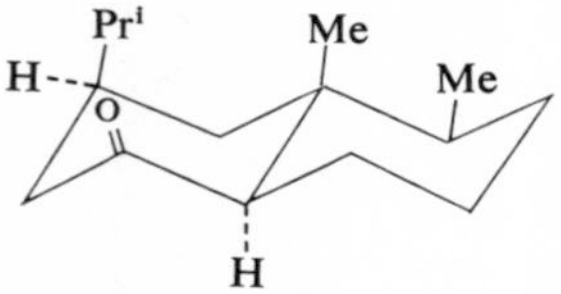

(27)

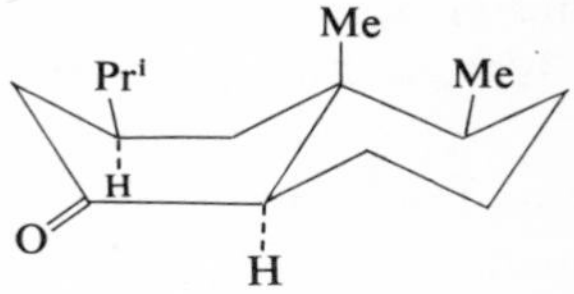

(28)

the trigonal $>$C$=$O centre also reduces the energy for chair→boat inversion.

In a *cis*-decalin there is the possibility of two alternative chair-chair conformations, e.g. (29) may adopt the 'steroid' conformation (30) or the 'non-steroid' alternative (31). *A priori* (31) should be preferred since

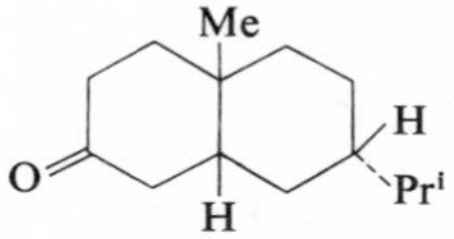

(29)

(30)   (31)

in (30) the i-Pr group is axially oriented.

9.5. Optical rotatory dispersion of decalones. Useful information regarding the conformations of decalones has been obtained by measurements of optical rotatory dispersion, i.e. of the manner in which the
optical rotation of an optically active form of e.g. (23), (26), or (29)
changes with changing wavelength (Djerassi, 1960; cf. Eliel, 1962).
This arises because conformational change: (24)→(25), (27)→(28), or
(30)→(31), alters the direction of the carbonyl dipole relative to the
bond vectors of the remainder of the asymmetric structure. This can be
seen by taking the carbonyl dipole as a reference axis, orienting the
structure in the same sense, e.g. as in (32) and (33), and viewing the
molecule from the carbonyl oxygen.

(30) ≡ (32)   (31) ≡ (33)

As the wavelength of measurement of optical rotation passes through
the region of carbonyl absorption (280–290 nm) the observed rotation is
found to change rapidly. With decreasing wavelength the rotation may
rise to a peak value and then fall, i.e. exhibit a positive Cotton effect,
or fall, passing through a trough, and then increase, i.e. exhibit a negative
Cotton effect. The sign of the Cotton effect may be correlated with the

spatial distribution of groups in (32) and (33) relative to three orthogonal planes, one through carbon atoms 3 and 10, one through carbon atoms 2, 3 and 4, and one through the $>CO$ group between carbon and oxygen, by what is known as the Octant rule (Djerassi, 1960; cf. Eliel, 1962).

9.6. Hydrindanes. These, like the decalins, exist in *cis* and *trans* forms e.g. (35), (36), (38). Simple hydrindones were obtained (Hückel, 1926; 1935) as follows by Claisen cyclisation and hydrolysis and decarboxylation of the intermediate β-keto ester. Since *cis*-hydrindan-1-one

cis and *trans* forms

(34)   (35)   (36)

cis or *trans* forms

(37)   (38)

(38) is more stable than the *trans* isomer (cf. chap. 2) *cis*-hydrindan-1-one is the only product of ring closure of the *cis*- or *trans*-ester (37) in the presence of alkoxide. Alkoxide would epimerise any *trans*-hydrindan-1-one formed, and also the *trans*-diester precursor (37). Hydrindane structures may also be obtained by general methods of ring contraction from a suitable decalin derivative, as in Woodward's cholesterol synthesis (Woodward *et al.*, 1952) and other examples (Johnson *et al.*, 1956), e.g.:

9.7. Bicyclo[3,3,0]octanes. *cis-* and *trans*-Bicyclo[3,3,0]octan-2-one
(40) and (41) were obtained as follows, by ketonisation of the cyclopen-
tane-1,2-diacetic acids (Barratt and Linstead, 1936).

cis or *trans* form

(39)   (40)   (41)

The *cis* form of cyclopentane-1,2-diacetic acid (39) was found to ketonise
much the more readily. This is reasonable in view of the strain in the
trans-bicyclo-octanone (41). The corresponding bicyclo-octanes obtained from the ketones by Wolff–Kishner reduction were found to
differ considerably in heat of combustion, the value for the *cis* isomer
being less than that for the *trans*, by some 25 kJ/mole. Thus *trans* fusion
of two cyclopentane rings represents a highly strained system.

The required cyclopentane-diacetic acids were synthesised as follows:

(i) $CH_2(CN)CO_2Et$/piperidine; (ii) reduction using Al/Hg in ether; (iii) HCl
hydrolysis.

9.8. Bicyclo[5,3,0]decanes. *cis-* and *trans-*Perhydroazalenes, or bicyclo-[5,3,0]decanes, (42) and (43), were synthesised in a similar manner to that used for the bicyclo-octanes (Allinger and Zalkow, 1961) viz.:

(i) LiAlH₄; PBr₃; KCN; (ii) Cyclisation using PhNMeLi in ether; (iii) Wolff–Kishner reduction.

The synthesis of the bicyclodecenone (46) which was the basis of the

(i) heat with ZnCl₂; (ii) O₃; (iii) aq. Na₂CO₃.

first synthesis of an azulene is interesting as an instance of transannular ring closure (St Pfau and Plattner, 1936). The acid catalysed migration of the olefinic bond during the dehydration of (44) so as to bring the olefinic bond into the most stable $\Delta^{9,10}$-position in (45) is also to be noted.

9.9. Perhydroanthracenes. Perhydroanthracenes are known in all the possible stereoisomeric forms, many of which are available by hydrogenation of anthracene under various conditions, e.g.:

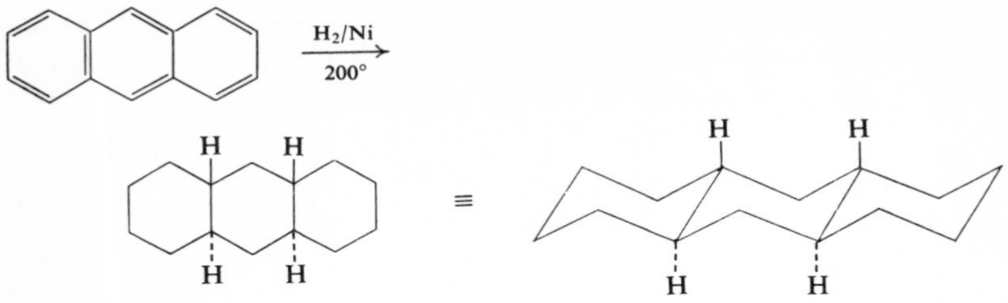

(47)

The Diels–Alder synthesis is also of interest; in (48) the dione containing

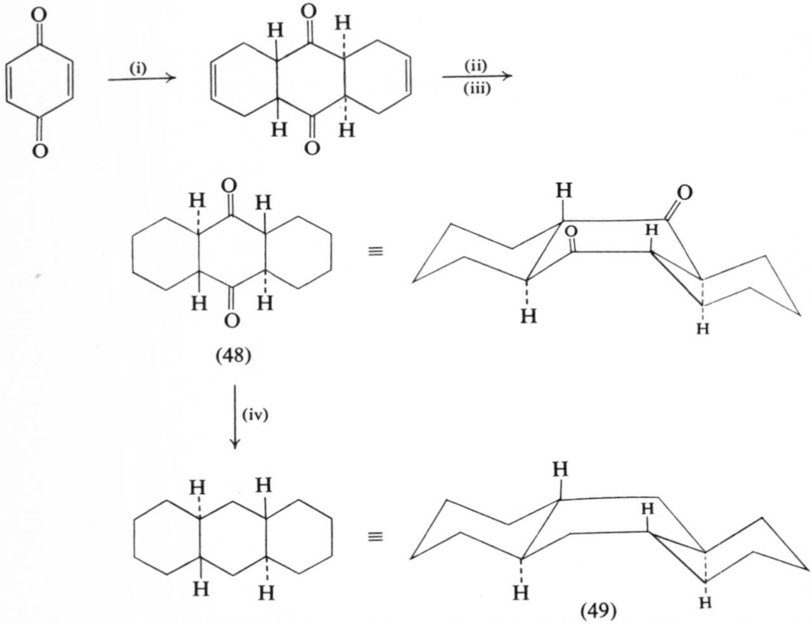

(48)

(49)

(i) butadiene; (ii) alkali inversion at the bridgeheads; (iii) reduce olefinic bonds;
(iv) Wolff–Kishner reduction.

ring adopts the twisted boat conformation as in cyclohexan-1,4-dione (cf. chap. 2), and reduction of the carbonyl groups then gives the *trans: anti: trans*-perhydroanthracene (49) with the chair-twisted boat-chair conformation as shown (Clarke, 1961).

The difference in heat of combustion of the *trans:syn:trans* (47) and *trans:anti:trans* (49) isomers, i.e. 21 kJ/mole, gives a value for the enthalpy difference between the chair and boat conformations for cyclohexane (Margrave *et al.*, 1963) (cf. chap. 2).

9.10. Decalins and bicyclo[5,3,0]decanes by transannular cyclisation.
The principle of transannular cyclisation indicated in the preparation of the bicyclodecenone (46) is illustrated in an interesting and wider context in relation to the following studies relating to sesquiterpene biosynthesis. Germacatriene (50), which contains a cyclodeca-*trans, trans*-diene ring is cyclised by a variety of electrophilic reagents, e.g. by acid to give

(50)

(51)   (52)

products (51) and (52) containing the ring structure characteristic of the selinane group of sesquiterpenes (Sutherland and Brown, 1968).

trans-Olefinic bonds located in medium-sized rings are somewhat strained by twisting of the ring (cf. chaps. 1 and 8), their relative reactivity depending on the degree of strain release as electrophilic addition occurs. In (50) cyclisation is directed by the greater reactivity of the 1,2- in comparison with the 5,6-double bond. However, the direction of ring closure may be altered by using the 5,6- or 1,2-epoxides (53) and (54), which on cyclisation gave (55) and (56), containing structures characteristic of the selinane and guaiane groups of sesquiterpenes respectively.

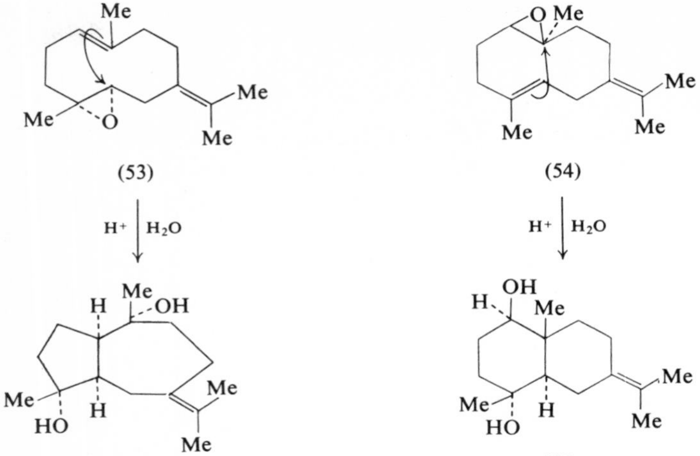

(53)   (54)

(56)   (55)

The principle of cyclisation of a hexa-1,3 *cis*,5-triene to a cyclohexa-1,3-diene, noted in chap. 4, provides an interesting means of interconverting a cyclodecatriene and a hexahydronaphthalene structure. This may be illustrated by the example of (57) which is photochemically cleaved to the cyclodecatriene (58) and this in turn may be thermally recyclised to give (59) and (60) (Corey and Hortman, 1963). The overall

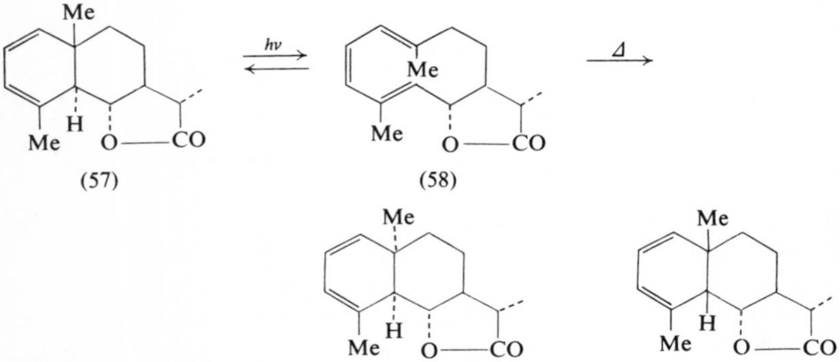

(57)   (58)

(59)   (60)

inversion: *trans* (57)→*cis* (59) and *cis* (60) arises because the photochemical step (57)→(58) is conrotatory and the thermal recyclisation is a disrotatory process (cf. chap. 4).

An interesting method of fragmentation which converts a decalin derivative into a cyclodecadiene may also be mentioned (Marshall and Bundy, 1966).

References

Allinger, N. L. and Zalkow, V. B. (1961). *J. Amer. Chem. Soc.*, **83**, 1144.

Barratt, J. W. and Linstead, R. P. (1936). *J. Chem. Soc.*, 611.

Corey, E. J. and Hortman, G. (1963). *J. Amer. Chem. Soc.*, **85**; cf. Sammes, P. G. (1970). *Quarterly Review*, **24**, 37.

Clarke, R. L. (1961). *J. Amer. Chem. Soc.*, **83**, 965.

Djerassi, C. (1960). *Optical Rotatory Dispersion*, McGraw-Hill.

Eliel, E. L. (1962). *Stereochemistry of Carbon Compounds*, McGraw-Hill, 398.

House, H. O. and Trost, B. M. (1963). *J. Org. Chem.*, **28**, 3362; (1965). *ibid.*, **30**, 1341.

House, H. O. (1965). *Modern Synthetic Reactions*, Benjamin, 210.

Hückel, W. (1925). *Annalen*, **441**, 1; (1926) *ibid.* **451**, 109; 132; (1935). *ibid.* **518**, 155; (1963). *ibid.* **666**, 30.

Johnson, W. S. (1968). *Accounts of Chemical Research*, **1**, 1.

Johnson, W. S., Bannister, B., Pappo, R. and Pike, J. E. (1956). *J. Amer. Chem. Soc.*, **78**, 6354.

Johnson, W. S., Korst, J. J., Clement, R. A. and Dutta, J. (1960). *J. Amer. Chem. Soc.*, **82**, 614.

Johnson, W. S., Wiedhaup, K., Brady, S. F. and Olsen, G. L. (1968). *J. Amer. Chem. Soc.*, **90**, 5277; 5279.

Malhotra, S. F., Moabley, D. F. and Johnson, F. (1967). *Chem. Comm.*, 448.

Marshall, J. A. and Bundy, G. L. (1966). *J. Amer. Chem. Soc.*, **88**, 4291.

Margrave, J. L., Frisch, M. A., Bautista, R. G. Clarke, R. L. and Johnson, W. S. (1963). *J. Amer. Chem. Soc.*, **85**, 546.

Prelog, V. (1950). *J. Chem. Soc.*, 420.

Robinson, D. L. and Theobald, D. W. (1967). *Quarterly Reviews*, **21**, 314.

Sutherland, J. K. and Brown, E. D. (1968). *Chem. Comm.*, 1060.

St Pfau, A. and Plattner, Pl. (1936). *Helv. Chim. Acta*, **19**, 858.

Van Tamelen, E. E., Storri, A., Hessler, E. J. and Schmitz, M. (1963). *J. Amer. Chem. Soc.*, **85**, 3295.

Woodward, R. B., Sondheimer, F., Taub, D., Heusler, K. and McLamore, W. M. (1952). *J. Amer. Chem. Soc.*, **74**, 4223.

Appendix

Physical properties of some alicyclic substances

	$(CH_2)_{n-1}\ CH_2$	$(CH_2)_{n-1}\ CHOH$	$(CH_2)_{n-1}\ CO$
n	b.p. (°C)	b.p. (°C)	b.p. (°C)
3	−33	103	—
4	11	125	98
5	49.5	139	129
6	80.8	161	155
7	117	185	179
8	147	103/22 mm	195
9	172	102/12 mm	95/12 mm
10	201	125/12 mm	107/13 mm

Infrared data: (cm^{-1})

	$(CH_2)_{n-1}\ CH_2$	$(CH_2)_{n-2}\ \overset{CH}{\underset{CH}{\|}}$		$(CH_2)_{n-1}\ CO$
n	ν_{CH}	ν_{CH}	$\nu_{C=C}$	ν_{CO}
3	3102, 3009	3076	1640	1815
4	2896, 2870	3048	1566	1788
5	2965, 2868	3061	1611	1746
6	2960, 2853	3024	1646	1715
7	—	—	1651	1703
8	—	—	—	1692

N.m.r. data:

	$(\dot{\mathrm{CH}}_2)_{n-1}$ CH_2	$(\dot{\mathrm{CH}}_2)_{n-2}\begin{matrix}\mathrm{CH}\\ \| \\ \mathrm{CH}\end{matrix}$	$(\dot{\mathrm{CH}}_2)_{n-1}$ CO
n	τ_{CH_2}	$\tau_{\mathrm{C=CH}}$	τ_{COCH_2}
3	9.78	2.99	8.35
4	8.04	4.05	6.97
5	8.49	4.40	7.94
6	8.57	4.41	7.78
7	8.47	—	7.62
8	8.47	4.46	7.70

Some bicyclic and polycyclic alicyclic substances

	m.p. °C
Bicyclo[2,2,1]heptane	87
Bicyclo[2,2,1]heptan-2-ol *exo-*	128
.. *endo-*	152
Bicyclo[2,2,1]heptan-2-one	92
Bicyclo[2,2,2]octane	173
Bicyclo[2,2,2]octan-2-ol	222
Adamantane	260
Adamantan-1-ol	288
Cubane	130
Bullvalene	96
Hexamethyl bicyclo[2,2,0]hexadiene i.e. hexamethyl Dewar benzene	7
Hexamethyl prismane	91
Twistane	164

Index